00-WSM-2

NAVIGATING THE TMDL PROCESS: LISTING AND DELISTING

by:

Kent W. Thornton, Ph.D.
FTN Associates, Ltd.

Paul L. Freedman, P.E., DEE
Limno-Tech, Inc.

2003

Water Environment Research Foundation

The Water Environment Research Foundation, a not-for-profit organization, funds and manages water quality research for its subscribers through a diverse public-private partnership between municipal utilities, corporations, academia, industry, and the federal government. WERF subscribers include municipal and regional water and wastewater utilities, industrial corporations, environmental engineering firms, and others that share a commitment to cost-effective water quality solutions. WERF is dedicated to advancing science and technology addressing water quality issues as they impact water resources, the atmosphere, the lands, and quality of life.

For more information, contact:
Water Environment Research Foundation
635 Slaters Lane, Suite 300
Alexandria, VA 22314
Tel: (703) 684-2470
Fax: (703) 299-0742
www.werf.org
werf@werf.org

This report was co-published by the following organizations. For non-Subscriber sales information, contact:

In the U.S.:
Water Environment Federation
601 Wythe Street
Alexandria, VA 22314
Tel: (800) 666-0206
Tel: (703) 684-2452
Fax: (703) 684-2492
www.wef.org
pubs@wef.org

International:
IWA Publishing
Alliance House, 12 Caxton Street
London SW1H 0QS, United Kingdom
Tel: +44 (0) 20 7654 5500
Fax: +44 (0) 20 7654 5555
www.iwapublishing.com
publications@iwap.co.uk

Library of Congress Catalog Card Number: 2003103358
Printed in the United States of America
WERF ISBN 1-893664-75-9 IWAP ISBN: 1-84339-663-7 WEF ISBN: : 1-57278-215-3

FTN Associates, Ltd.
Limno-Tech, Inc.
Ross & Associates Environmental Consulting, Ltd.
Technology Planning and Management Corporation

ACKNOWLEDGMENTS

The project team would like to acknowledge and thank the states of Arkansas, California, Indiana, Louisiana, Massachusetts, Michigan, Mississippi, Oregon, and Rhode Island for their support and participation in the surveys for this project.

Report Preparation

Principal Investigators:

Kent W. Thornton, Ph.D.
FTN Associates, Ltd.

Paul L. Freedman, P.E., DEE
Limno-Tech, Inc.

Project Team:

Patrick Downey
FTN Associates, Ltd.

David Dilks, Ph.D.
Wendy M. Larson
Troy Naperala, P.E.
Daniel Schechter, P.E.
Limno-Tech, Inc.

Martha G. Prothro
Anne Dettlebach
Gerald W. Boese
Lane Nothman
Ross & Associates Environmental Consulting, Ltd.

Jose A.H. Sobrinho
Kristyn B. Stevens
Technology Planning and Management Corporation

Project Subcommittee

Norman E. LeBlanc (Co-Chair)
Hampton Roads Sanitation District

Philip B. Dorn, Ph.D. (Co-Chair)
Shell Global Services

Fredric P. Andes
Barnes & Thornburg

Donald J. Brady, Ph.D.
U.S. Environmental Protection Agency

Betsy A. Cooper
King County (Washington) Department of Natural Resources and Parks

Robbin W. Finch
Boise City (Idaho) Public Works

Michael Haire
U.S. Environmental Protection Agency

John D. Morton
ALCOA

Kenneth H. Reckhow, Ph.D.
Duke University; Water Resources Research Institute, North Carolina State University

David Roberts
Washington Department of Ecology

Jerald L. Schnoor, Ph.D., P.E., DEE
University of Iowa

Richard F. Schwer
DuPont Company

Thomas C. Stiles
Bureau of Water, Kansas Department of Health and Environment

Water Environment Research Foundation Staff

Project Manager: Dean E. Carpenter, D.Env.

ABSTRACT AND BENEFITS

Abstract:
The total maximum daily load (TMDL) program may have attracted as much attention as any program established by the U.S. Environmental Protection Agency (EPA) in its history. More than 34,000 comments were received and reviewed by EPA on the proposed changes in the TMDL rule alone. As a result of the controversy surrounding this program, the Water Environment Research Foundation initiated a project entitled *Navigating the TMDL Process: Listing and Delisting* and a companion project *Navigating the TMDL Process: Evaluation and Improvements.* This report is on the listing and delisting project.

The TMDL process is initiated by the listing process, which involves placing a water body on the 303(d) list. A water body is placed on the 303(d) list when it is impaired (not attaining water quality standards (WQS). Equally important is how a water body is delisted, that is, removed from the list, once the WQS are attained. This report provides a minimum set of activities that should be undertaken by the states, an investigation of alternatives and options, and a comprehensive approach to the listing and delisting processes that will continue to protect and sustain the quality of our nation's waters. Also addressed herein are ways to address and reduce the uncertainty in misclassification of waters that need TMDL studies.

The listing and delisting options and alternatives emerged from surveys of eight states, reviews of approaches used by selected other states, published literature, and from project team experience. These alternatives can reduce uncertainty in each of the major elements of the listing and delisting process: (1) WQS (designated uses and water quality criteria); (2) water quality data acquisition; (3) comparative methodology; and (4) list content and format. Reducing uncertainty increases the likelihood that only impaired waters will be added the 303(d) list and that waters attaining WQS will be delisted.

The TMDL program is an important part of the effort to protect and sustain the waters of the US. These listing and delisting alternatives should assist states as they navigate the TMDL process.

Benefits:

- Recommends ways to improve the listing and delisting process based upon surveys of eight states and discussions with other selected states;
- Evaluates alternatives and options for reducing uncertainty for each of the four major elements of the listing/delisting process; and
- Proposes a set of minimum activities, a number of alternatives and options, and a comprehensive approach for reducing uncertainty in the listing and delisting process.

Key Words: water quality standards, designated use, impairment, nonattainment, 303(d)

CONTENTS

LIST OF TABLES

LIST OF FIGURES

LIST OF ACRONYMS

AFBF	American Farm Bureau Federation
AMSA	Association of Metropolitan Sewerage Agencies
ASIWPCA	Association of State and Interstate Water Pollution Control Administrators
AU	Assessment unit
BMP	Best management practice
CFR	Code of Federal Regulations
CPP	Continuous planning process
CSO	Coastal States Organization
CWA	Clean Water Act
DO	Dissolved oxygen
ECOS	Environmental Council of the States
EPA	U.S. Environmental Protection Agency
FACA	Federal Advisory Committee Act
FWQC	Federal Water Quality Coalition
FY	Fiscal year
GAO	General Accounting Office
IBI	Index of biotic integrity
MCL	Maximum contaminant level
NPCC	National Pork Producers Council
NPDES	National Pollution Discharge Elimination System
NRC	National Research Council
NRDC	Natural Resource Defense Council
NWEA	Northwest Environmental Advocates
QA	Quality assurance
QA/QC	Quality assurance/quality control
TMDL	Total maximum daily load
UAA	Use attainability analysis
WEF	Water Environment Federation
WERF	Water Environment Research Foundation
WQS	Water quality standards

EXECUTIVE SUMMARY

The total maximum daily load (TMDL) process is initiated when a water body is placed on the 303(d) list (listing process). Because of the interest in the entire TMDL program, the Water Environment Research Foundation (WERF) initiated studies to guide its subscribers and other water quality professionals as they navigate the TMDL process. This project focused specifically on the listing and delisting process. A companion WERF project will focus on evaluation and improvements of the TMDL process.

This project had two objectives:

- To develop a practical, reasonable, and scientifically defensible process for listing and delisting water bodies, in accordance with the governing statute, based on review of a range of state programs and technical approaches, recommendations from selected stakeholders and from the TMDL committee convened under the Federal Advisory Committee Act (FACA), and the project team's experience with various aspects of the TMDL process and interpretation of water quality information; and
- To evaluate the TMDL process in selected studies for its scientific and technical merit and its ease of implementation and interpretation.

To achieve the project objectives, the project team reviewed comments submitted by the Water Environment Federation, Association of State and Interstate Water Pollution Control Administrators, Association of Metropolitan Sewerage Agencies, TMDL Coalition, and other selected organizations on the proposed TMDL rule. In addition, the team reviewed the FACA committee's final report to the United States Environmental Protection Agency (EPA) and conducted a survey of paired, contiguous states to determine the current approaches used for listing and delisting, as well as to request additional information on the TMDL process that would be used for the companion WERF project. Finally, the team evaluated these data and proposed listing and delisting alternatives.

The listing process encompasses four major components: applicable water quality standards (WQS), the water quality data compiled to compare with the WQS, the comparison methodology used to determine WQS impairment, and the format and organization of the 303(d) list. The survey carried out for this project indicated that there was considerable variability as to how the states integrate these four components into the listing process. This variability results in uncertainty in the appropriateness of some listing decisions.

Based on the evaluation of listing and delisting alternatives reviewed as part of this study, the project team proposed three recommendations for reducing uncertainty associated with each of the four components listed above:

- A minimum set of activities should be considered by all states as part of the listing and delisting process. These activities should include (1) a critical review of WQS, including revisions since the last assessment as well as revisions proposed for the next triennial review or legislative session; (2) establishment of quality assurance/quality control (QA/QC) requirements for all data used as part of the listing and delisting process; development of a statistically based comparison methodology that explicitly considers Type I vs. Type II errors; (3) specification of the confidence level (error rate) desired (permitted) in the tests comparing data with the WQS; and (4) adoption of List Content and Format categories in *EPA 2002 Integrated Water Quality Monitoring and Assessment Report Guidance* ("EPA Listing Guidance").

- Consideration of options and alternatives that various states have implemented to reduce uncertainty in WQS (e.g., tiered WQS, use attainability analysis); water quality data acquisition (e.g., third-party QA/QC, protocols); comparative methodologies (e.g., binomial statistical tests, minimum sample sizes); List Content and Format categories (e.g., EPA Listing Guidance, Florida Option for Delisting). Figure ES-1 highlights some options and alternatives that should be considered as part of this effort.

- Adoption of a comprehensive approach by implementing as many of these alternatives as state policies, time, and resources permit.

The TMDL program is an important part of the effort to protect and sustain the waters of the United States. These listing and delisting alternatives should assist subscribers and other water quality professionals as they navigate the TMDL process.

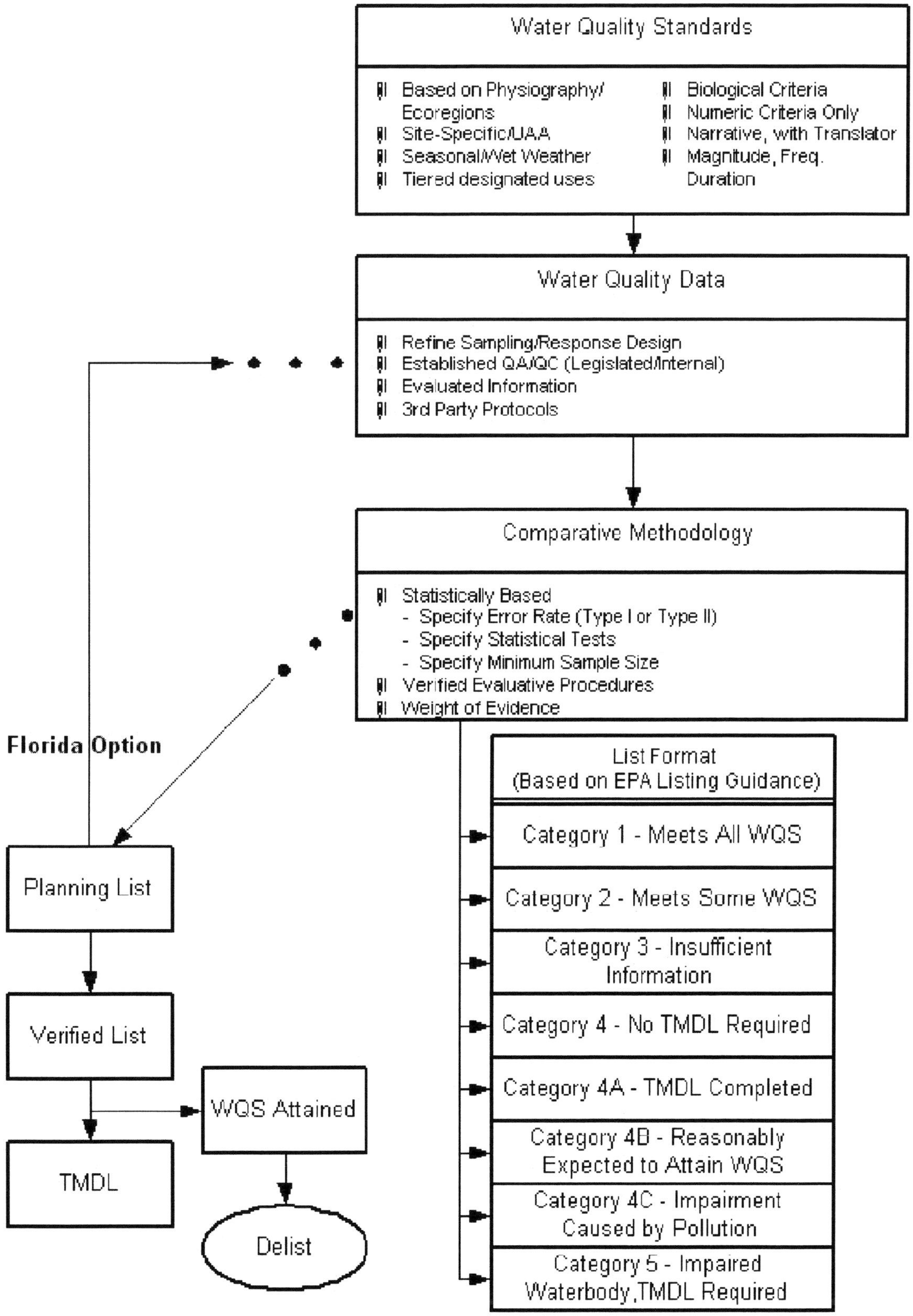

Figure ES-1. Four Major Elements of the Listing/Delisting Process and Options or Alternatives for Reducing Uncertainty and Navigating through the Process (after GAO 2002).

CHAPTER 1.0

INTRODUCTION

1.1 Purpose

The total maximum daily load (TMDL) program may have attracted as much attention as any program established by the U.S. Environmental Protection Agency (EPA) in its history. The TMDL program arose from Section 303(d), a relatively obscure subsection of the 1972 Federal Water Pollution Control Act Amendment (supplemented by the 1977 Clean Water Act and the 1987 Water Quality Act). Over the past several years, the TMDL program has gained prominence as a driving force behind many aspects of EPA's water programs. Opinions about the TMDL program range from it being the key to protecting, managing, and restoring water bodies throughout the United States, to it exerting unauthorized control over public and private land use.

The TMDL process begins when a water body is placed on the 303(d) list. Listing indicates the water body or water body segment is not attaining applicable water quality standards (WQS). Once a TMDL is developed, the water body or segment is delisted (removed from the 303(d) list). Because listing initiates and delisting concludes the TMDL process for waters, the Water Environment Research Foundation (WERF) and its subscribers were interested in reviewing the science underlying listing and delisting processes and decisions.

The purpose of this report is to present research results on alternative scientific approaches for listing and delisting waters. The audience for this report is water quality professionals — stakeholders, the states, and EPA.

1.2 Background

Under Section 303(d) of the Clean Water Act, states are required to identify waters not attaining WQS even after point sources of pollution have installed the minimum required levels of pollution control technology. States must also establish priority rankings for waters on the list and develop TMDLs for each listed water body. A TMDL specifies the maximum amount of a pollutant that a water body can assimilate and still meet WQS, and subsequently allocates this maximum pollutant load among point and nonpoint sources with a margin of safety. EPA must approve or disapprove state lists and TMDLs. If a state submission is inadequate, EPA must establish the list or the TMDL (EPA 1999).

In 1996, the EPA Office of Water determined there was a need for a comprehensive evaluation of EPA's and the states' implementation of their 303(d) responsibilities. EPA convened a committee under the Federal Advisory Committee Act (FACA) on the TMDL program to undertake this evaluation and make recommendations for improving the process, including the TMDL regulations and guidance. In 1998, the FACA TMDL Committee provided its final report with more than 100 recommendations focused on the listing and delisting process, TMDL development, scicncc and tools, and TMDL program management (EPA 1998a). This report and its recommendations, along with other comments, guided the development of proposed EPA revisions to the TMDL process (EPA 1999).

In 1999, EPA announced these proposed revisions in the Federal Register and solicited comments from the public on the revisions. EPA received and reviewed more than 34,000 comments from a variety of different industries, municipalities, trade organizations, professional societies, nonprofit organizations, states, and other federal agencies on these proposed regulations. EPA considered these comments, modified its proposal in some respects, and issued the final rule for the TMDL program on July 13, 2000 (65FR 43856). The promulgation of this rule generated additional controversy in terms of both the rule content and the process through which it was promulgated. Because of the continued controversy, Congress prohibited EPA from implementing the final rule through a spending prohibition attached to the Military Construction Appropriations Act: FY 2000 Supplemental Appropriations (PL106-426) on July 14, 2000. The Act prohibited EPA from spending funds on the TMDL rule until October 2001. On August 9, 2001, EPA announced it would delay the effective date of implementing the TMDL rule for 18 months to permit the agency to solicit additional comments, review the June 2001 National Research Council report, *Assessing the TMDL Approach to Water Quality Management*, and revise the TMDL program to improve its effectiveness and flexibility. In the meantime, the TMDL program continues to operate under the rules promulgated in 1985, as amended in 1992.

1.3 Project Objectives

In full consideration of the controversy and uncertainty associated with the TMDL rules, the objectives of this project were:

- To develop a practical, reasonable, and scientifically defensible process for listing and delisting water bodies, in accordance with the governing statute, based on review of a range of state programs and technical approaches, recommendations from selected stakeholders and from the TMDL committee convened under FACA, and the project team's experience with various aspects of the TMDL process and interpretation of water quality information; and
- To evaluate the TMDL process in selected studies for its scientific and technical merit and its ease of implementation and interpretation.

Alternative approaches were then evaluated for their ease of implementation and interpretation. This project emphasizes the listing and delisting processes. A WERF companion project, *Navigating the TMDL Process: Evaluation and Improvement,* is addressing issues associated with the estimation and allocation of pollutant loads.

1.4 General Approach

Per the project scope of work, the project team addressed the two objectives by reviewing comments on the TMDL Program and proposed rule-making. In addition, the team reviewed listing and delisting processes, including the monitoring programs, used in selected but representative states. After synthesizing and summarizing the advantages and disadvantages of the programs based on scientific and technical principles and public policy and programmatic concerns, the team was able to formulate a process that is scientifically defensible, reasonable and workable, and that is applicable or adaptable for many type of aquatic systems and various types and degrees of impairments.

1.5 Report Organization

The first chapter of this report provides background on the TMDL program and issues, the project objectives, and general approach. The second chapter provides detailed background on the TMDL program and process. The third chapter provides a synthesis of some of the review comments on the listing and delisting process that were provided to EPA in response to both the proposed and the 2000 TMDL rule. The fourth chapter discusses the survey that was designed to solicit information from the states on the listing and delisting process, the results from the survey, and reviews of other selected state approaches. The fifth chapter builds on the results and recommendations from the state surveys, the experience of the project team, and recommendations provided both in the review comments to EPA on the TMDL rule and the NRC report recently submitted to U.S. Congress on the role of the TMDL process in water quality management. This chapter proposes a set of minimum activities, alternatives and options, and a comprehensive approach to reducing uncertainty in the listing-delisting process. These alternatives and options relate to the four major elements of the listing-delisting process: WQS, water quality data acquisition, comparative methodology, and list content and format. The final chapter provides the conclusions and recommendations from the study. References and appendices follow these six chapters.

CHAPTER 2.0

BACKGROUND ON TMDL PROCESS

This background is included to provide a framework for the rest of the document. Much of it is taken from existing EPA reports.

2.1 1972 Federal Water Pollution Control Act Amendments

The Federal Water Pollution Control Act Amendments of 1972 (PL 92-500), as supplemented by the Clean Water Act of 1977 and the Water Quality Act of 1987, are the foundation for protecting the nation's water resources. The overarching goal established by the Federal Water Pollution Control Act Amendments was that of restoring and maintaining the "chemical, physical, and biological integrity of the Nation's waters."

Prior to the 1972 Act, the focus was on a "water quality-based" approach to pollution control that required states to develop WQS specifying the level of water quality each water body should achieve or maintain. Point source dischargers were then expected to comply with those WQS. This regulatory approach proved to be ineffective, in large part, because both compliance and enforcement depended on being able to prove a causal connection between the water quality impairment and a particular discharge.

The 1972 Water Pollution Control Act Amendments introduced (and to a large extent focused discussion on) a predominantly technology-based approach: all point source dischargers of certain pollutants were expected to limit their discharges by meeting nationally established effluent limitations. Effluent limitations were specified in National Pollution Discharge Elimination System (NPDES) permits, issued by federally approved states or by EPA to certain pollutant sources. State permits are subject to veto by EPA. Effluent limitation guidelines were set at a national level based on available technologies for point-source wastewater treatment appropriate to specific industry categories. Municipal treatment plants were also required to achieve specific control levels based on available technology. Applying and enforcing effluent limitations based on technology eliminated the need to link some required reductions at particular sources with the ambient condition of a water body (NRC 2001).

However, the Federal Water Pollution Control Act included a second tier of protection for the nation's waters. Under section 303 of the Act, states adopted WQS for their waters based, in part, on national criteria guidance from EPA. These WQS established the "designated use" of the water and included both general, narrative criteria and numeric, pollutant-specific criteria to protect the designated use. Water quality-based effluent limits were to be required in individual

NPDES permits where technology-based effluent limits were insufficient for the water body to meet water quality goals. Where the required technology- or water quality-based pollution controls applied to point sources of water pollution did not result in the attainment of the WQS, the water body was considered impaired.

Section 303(d) of the Act called for states to identify impaired waters and established the TMDL process to provide for more stringent water quality-based controls when technology-based controls were inadequate to achieve state WQS. A TMDL specifies the amount of a particular pollutant that may be introduced into a water body and allocates the total allowable pollutant loads among sources, thereby providing a plan to attain and maintain state WQS. TMDLs are established for water bodies impaired by point sources, nonpoint sources, or both; however, only point sources are required (through the NPDES permit program) to meet their TMDL load allocations. (The TMDL process is described in greater detail in section 2.3.)

The Federal Water Pollution Control Act Amendments and supplementary Clean Water and Water Quality Acts included several other programs that also helped restore and maintain the quality of the nation's water resources. These programs include national water quality criteria guidance, state nonpoint source management programs, funding provisions for municipal wastewater treatment facilities, and state water quality monitoring programs. All these programs, taken together, have produced significant and widespread improvements in water quality over the last quarter-century.

2.2 Listing and Delisting

Section 303(d) of the Clean Water Act directs states, authorized tribes, and/or EPA to identify and list all waters for which the technology-based standards are not stringent enough to meet applicable WQS. EPA regulations at 40 CFR 130.7(b)(5) directs states to base listing decisions on "all existing and readily available water quality-related data and information," including both monitored and evaluated data and information. "Monitored" data refers to direct measurements of water quality, including sediment and some fish tissue analyses. "Evaluated" data and/or information provide an indirect appraisal of water quality through information on historical adjacent land uses, riparian health and habitat, the location of sources, results from predictive modeling using input variables, some surveys of fish and wildlife, and citizen complaints.

The Clean Water Act does not directly address the issue of removing waters from the 303(d) list. According to EPA's guidance for the 1994 section 303(d)(1) list, states may remove waters when: (1) new information shows that "the original basis for listing is determined to be inaccurate" or (2) EPA has approved a TMDL designed to attain WQS.

2.3 Total Maximum Daily Load (TMDL) Process

Generally, a TMDL is a means for recommending the level of control/reduction needed to meet WQS in a particular water body or watershed. TMDLs contain specific goals for a given water body or watershed and allocate load control/reduction responsibilities among contributing sources. Establishing a TMDL is a very important step in watershed protection because it quantifies goals for water quality conditions that may then determine what actions are needed to restore or protect the health of the water body.

Under section 303(d), states are required to identify and establish a priority ranking for waters not meeting WQS, taking into account the severity of the pollution and the uses to be made of the waters. EPA is required to review each state list and approve it or, if it is deemed inadequate, to disapprove it and prepare a list for the state. Once the list is established, states are to develop a TMDL for each constituent not meeting WQS for each listed water body.

In 1992, EPA amended its regulations to describe in greater detail requirements for states to submit lists of waters needing TMDLs. Among other things, the revised regulations required states to submit lists every two years and to target waters for which TMDLs would be developed during the next two years. In 1997, EPA issued a policy memorandum providing that states should generally set 8- to 13-year overall schedules for developing TMDLs for waters listed under section 303(d)(1) (beginning with the 1998 list). The memorandum included a set of factors that states can use to set longer schedules in exceptional cases.

By statute, EPA is required to review and approve or disapprove each TMDL (within 30 days of the submittal by the state) and then establish a TMDL in the case of any disapproval. Section 303(d) provided that TMDLs were to be developed "at a level necessary to implement the applicable WQS with seasonal variations and a margin of safety which takes into account any lack of knowledge concerning the relationship between effluent limitations and water quality."

Once a TMDL has been established for a water body (or segment) and the appropriate source loads developed, control actions are to be implemented through permits for point sources. States are generally expected to implement nonpoint source management programs consistent with their overall planning process, including TMDLs. The means for achieving specific nonpoint source load allocations is addressed in Section 303(e) and 319 of the Clean Water Act. According to EPA guidance, monitoring is also a crucial element of the TMDL process that allows the TMDL's effectiveness in achieving attainment of WQS to be evaluated.

2.4 Federal Litigation

Although the TMDL provisions of section 303(d) of the Clean Water Act were enacted in 1972, states and EPA did not emphasize their implementation until quite recently. As noted above, the initial focus of state and EPA efforts was to establish performance standards for industrial dischargers and sewage treatment facilities. As a result, many state lists of waters needing a TMDL were limited to a few water bodies, and few TMDLs were developed. State and EPA permit writers developed wasteload allocations for their point sources to meet WQS, but this was often done without a listing of impaired water bodies or a TMDL process that met the specific terms of section 303(d).

The Clean Water Act specifies that EPA has a duty to review and approve the list of impaired water bodies and TMDLs submitted by a state and to develop a list and/or TMDL where a state submission was disapproved. Because the Act did not establish a schedule or deadline for states to submit lists and TMDLs, some states did not develop lists or TMDLs, and neither did EPA take a disapproval action or step in to develop lists or TMDLs for those states.

In the early 1990s, citizen organizations began bringing legal actions against EPA, seeking the listing of water bodies and establishment of TMDLs. To date, environmental groups have filed about 41 legal actions in 42 states. Most of the lawsuits have been filed since the mid-1990s. They generally argue that EPA has failed to carry out its mandatory duty to disapprove inadequate state 303(d) lists or to develop such 303(d) lists and TMDLs where the state had failed to do so. More than 20 of these lawsuits have resulted in court orders or consent decrees under which EPA is required to establish TMDLs if the state fails to do so in a timely manner.

2.5 Recent EPA-Promulgated Rule-Making

EPA's existing regulations implementing the listing and TMDL processes took their current form in 1985 and were amended in 1992. These rules are found primarily in 40 CFR 130.7, with related rules in 40 CFR 122.

By the mid-1990s, EPA had initiated a new effort with states and other stakeholders to put the TMDL program on a sounder footing. At the same time, EPA developed a clearer standard for determining the sufficiency of state implementation of a TMDL program with requirements that: (1) the state has established TMDLs; (2) the state has a plan for implementing TMDLs in the future; and (3) the state is engaged in and supporting the plan.

In 1996, EPA determined that there was a need for a comprehensive evaluation of EPA's and the states' implementation of their section 303(d) responsibilities. EPA convened a committee under the Federal Advisory Committee Act (TMDL FACA committee) to undertake this evaluation and make recommendations for improving implementation of the TMDL or Section 303(d) program, including recommended changes to the TMDL regulations and guidance. The TMDL FACA committee comprised 20 individuals with diverse backgrounds in agriculture, forestry, environmental advocacy, industry, and state, local, and tribal governments. In 1998, the committee submitted its final report to EPA containing more than 100 consensus recommendations, a subset of which would require regulatory changes to be implemented.

Building on the recommendations of the TMDL FACA Committee, EPA drafted revisions to the existing TMDL regulations and published proposed TMDL program changes in the August 1999 *Federal Register*. EPA engaged in an extensive outreach and information-sharing effort following the publication of the proposed rule. More than 34,000 comments were received from industries, municipalities, trade organizations, professional societies, state agencies, nonprofit organizations, and federal agencies. Some comments are summarized in Chapter 3.0.

In July 2000, EPA published final revisions to the TMDL regulations. Key elements related to listing and delisting processes included:

- More comprehensive lists of all water bodies that do not attain and maintain state WQS with revision of the lists every 4 years rather than every 2 years;
- A four-part 303(d) list, of which only a subset would have TMDLs developed;
- A schedule, based on stated priority factors, for establishing all needed TMDLs over a period of 10 years, with an allowance for another 5 years where necessary;
- Expanded opportunities for public comment on listing methodologies, lists, prioritized schedules and TMDLs prior to submission to EPA; and
- Delisting, without a TMDL if new data or information indicate that the water body is attaining and maintaining applicable WQS for the pollutant.

As EPA was finalizing these rules, Congress adopted an amendment to the Military Construction Appropriations Act: Supplemental Appropriations (PL 106-426) to stop the implementation of the rule. This provision prohibited EPA from using FY 2000 or 2001 funds to "make a final determination on or implement" the July 2000 regulations. This action was taken in response to continued controversy over the rule. EPA published the final rule before the legislation was enacted but established the effective date of the new TMDL regulations as 30 days after the expiration of the prohibition, if enacted, or at the earliest, October 30, 2001. More recently, on August 9, 2001, EPA published a notice in the *Federal Register* proposing to delay by 18 months the effective date of the TMDL rules, giving an effective date of April 2003.

2.6 Recent EPA Guidance

EPA has provided recent guidance and refinement on the process for assessing and listing water bodies. EPA's 2002 Integrated Water Quality Monitoring and Assessment Report Guidance (EPA 2001b) will be referred to elsewhere in this document. In addition, EPA has released its Consolidated Assessment and Listing Methodology (CALM; EPA 2001c). These two guidance documents are summarized below.

2.6.1 2002 Integrated Water Quality Monitoring and Assessment Report Guidance

The stated purpose of this guidance ("EPA Listing Guidance") is to provide states, territories, and authorized tribes with guidance for "integrating the development and submission of 2002 305(b) water quality reports and Section 303(d) lists of impaired waters" (EPA 2001b). A major provision of the guidance is a description of five unique categories for identifying the water quality attainment status of water bodies. These five categories are further described in Section 5.2.4 and Figure 5-2 of this document where format options and alternatives for 303(d) lists are discussed.

EPA Listing Guidance did not change the rules for listing and delisting. It did, however, provide clarification for identifying waters requiring a TMDL and provided a framework for combining the 305(b) report and the 303(d) list into an Integrated Report. The guidance identified the following components that the Integrated Report should include:

- An assessment methodology;
- The delineation of assessment units based in the National Hydrography Dataset and an integrated list of all the assessment units in the state or territory in the five categories described in the guidance;
- Data and information supporting the categorization of each assessment unit;
- A description of the public participation process and a summary of the comments received and the responses made to the comments; and
- An assessment report based on a probability design if such a design is a component of the monitoring strategy (EPA 2001b).
- In March 2002, EPA issued additional clarification guidance to address the use of biological data in the Integrated Report (EPA 2002).

2.6.2 Consolidated Assessment and Listing Methodology (CALM) Guidance

The stated objective of the CALM guidance (EPA 2001c) is to "provide a framework for states, territories, interstate commissions, and authorized tribes to document the decision making process used to assess WQS attainment." The guidance is presented as a series of questions to be answered in order to document methodology. The document is organized into three parts. Part A discusses how different types of data (e.g., chemical, biological, toxicity) can be used to support

WQS attainment decisions. Part B addresses the design of monitoring programs, data quality objectives, and the amount and scale of data needed. Part C describes approaches for 305(b) and 303(d) reporting.

CHAPTER 3.0

SUMMARY OF COMMENTS ON PROPOSED RULE

3.1 Overview

On August 23, 1999, EPA issued proposed revisions (64 Fed. Reg. 46012-46089; August 23, 1999) to the Water Quality Planning and Management Regulation (40 CFR Part 122). The Final Rule for the regulations was issued on July 13, 2000. EPA has since withdrawn the Final Rule pending further review and comment. The preamble of the Final Rule included a discussion of comments received by EPA on various issues and EPA's response to those comments in producing the Final Rule.

This chapter provides a summary of comments on the Final Rule submitted by certain professional associations and organizations. The comments were reviewed to gain additional insight into listing and delisting issues that were repeatedly raised by water quality professionals. Understanding the issue or problem is the first step toward solving it. The organizations were selected because their interests, membership, or perspectives were similar to those of the WERF membership. This summary is based on a review of the comments as submitted by the interested parties rather than the summary of comments published by EPA with the Final Rule. Section 3.2 provides a narrative summary of comments provided by the professional associations and organizations listed in Table 3-1. Comments are not attributed to individual sources.

Section 3.3 provides a summary of some of the issues raised at the 2001 TMDL Science Conference in St. Louis. Nearly 600 individuals participated in the conference and represented a broad array of stakeholders. Section 3.4 discusses the results from two congressionally authorized reviews, one by the General Accounting Office (GAO), and the other the National Research Council's report, *Assessing the TMDL Approach to Water Quality Management* (NRC 2001). Section 3.5 summarizes a review of the EPA Listing Guidance prepared by the Federal Water Quality Coalition, American Farm Bureau, and Association of Metropolitan Sewerage Agencies.

Table 3-1. **Professional Organization and Association Comments on the Final Rule Reviewed for This Project. Members of the listed organizations or associations are indicated, where appropriate.**

Name of Organization	
TMDL Coalition comprised of:	
Alcoa, Inc.	Edison Electric Institute
Alliance of Automobile Manufacturers	Gary (Indiana) Sanitary District
American Forest and Paper Association	General Electric Company
American Gas Association	Mid America Crop Protection Association
American Iron and Steel Institute	Minnesota Power
American Petroleum Institute	National Association of Homebuilders
American Chemistry Council (formerly Chemical Manufacturers Association)	National Mining Association
City of Hamilton, Ohio	Pharmaceutical Research and Manufacturers of America
City of Indianapolis, Indiana	Rubber Manufacturers Association
City of Lima, Ohio	Western States Petroleum Association
Joint comments by: Association of State and Interstate Water Pollution Control Administrators (ASIWPCA) Environmental Council of the States (ECOS) Coastal States Organization (CSO)	
Association of Metropolitan Sewerage Agencies (AMSA)	
Water Environment Federation (WEF)	
National Pork Producers Council (NPPC)	
Natural Resources Defense Council (NRDC)	
Northwest Environmental Advocates (NWEA)	

3.2 Summary of Comments Provided by Professional Associations and Organizations

Comments provided by professional associations and organizations are summarized in Table 3-2. The comments are organized according to four primary program components relating to the listing process (WQS; assessment data; assessment methodology; and list content, format and submission. In addition, some comments fall into the categories of delisting and general comments.

The purpose of this summary is not to identify comments with particular entities, but rather to indicate concern of selected interested parties over the listing process. Issues and comments indicated in Table 3-2 were raised by one or more of the entities listed in Table 3-1.

Table 3-2. Professional Association and Organization Comments Summary on Selected Aspects of the Proposed Rule Pertaining to Listing and Delisting of Water Bodies

Component of Listing Process	Range of Comments
WQS	1. Narrative criteria are insufficient to support listing decisions without supporting data from quantitative criteria. 2. Biological criteria are insufficient to support listing decisions without supporting data from quantitative criteria 3. List waters failing to meet WQS for any reason. 4. Nonattainment of WQS due to unknown causes is not a valid reason for listing. 5. Information that is not clearly part of a state's water quality criteria (e.g., fish consumption advisories, drinking water MCLs) should not be used to list water bodies. 6. Water bodies should meet WQS at all times in all places. 7. Listing decisions should be made based on quantitative evidence of impairment.
Assessment Data	1. Listing decisions should be based on current data that meet specified data quality objectives and analytical standards. 2. Older data are not necessarily less valid than more recently collected data. 3. There should be national consistency in defining data quality (e.g., data quality requirements no less stringent than those required for NPDES compliance determinations) and quantity. 4. EPA should clearly define "impairment" and address issues of minimum number, type, and duration of samples.
Assessment Methodology	1. Listing should be based only on the exceedance of a WQS. Clearly define "attainment" of WQS and "impairment." 2. Do not list for classes of pollutants (e.g., metals, pesticides). 3. Methodology should be developed as part of the continuous planning process (CPP) instead of the 303(d) process. 4. Methodology should be scientifically defensible. 5. EPA should outline listing criteria that apply to all states.
List Content, Format, Submission	1. No 4-part list. 2. Pollutants should be specifically identified (Don't list for unknown causes). 3. List cycle should be 5 years. 4. Retain 2-year cycle for at least two more cycles before considering moving to a 4- or 5-year cycle. 5. Don't include water bodies that meet WQS but show trends toward declining water quality (i.e., threatened waters). 6. Include threatened and "expected to meet" waters. 7. List all waters failing to meet WQS for any reason.
Delisting	1. States should clearly articulate process. 2. Process should be no more complex or stringent than listing process. 3. Delist when WQS have been attained. 4. Delist based on credible data.
General Comments	1. Encourage watershed planning alternatives to development of TMDLs. 2. Compliance with load allocations and wasteload allocations should be verified, not just assumed. 3. Establish alternative listing process taking into account waters that a. cannot be classified as "pollutant" impaired b. are expected to attain WQS c. have developed TMDLs 4. Rule could result in further implementation delays and unenforceability of TMDL program.

3.2.1 WQS

Although WQS review is not part of the listing or TMDL process, the decision to list a water body depends on whether it meets WQS. Therefore, comments related to WQS were included as part of this review. State WQS are made up of designated uses, water quality criteria that support designated uses, and an antidegradation policy. Some comments stressed the importance of emphasizing scientific rigor in the WQS program to ensure that both designated uses and water quality criteria for individual sites are appropriate and attainable. Several comments addressed the difficulty of applying certain WQS in a determination of attainment or nonattainment. Specifically, all the organizations indicated that narrative criteria including fish consumption advisories should not be used to list water bodies unless there were objective, quantifiable, supporting evidence of impairment. One organization also commented that biological criteria, including fish consumption advisories, should not be used as the sole basis for listing. One comment raised concern on how the antidegradation portion of the WQS would be applied. The Final Rule states that threatened waters need not be included in the 303(d) list.

3.2.2 Assessment Data

The actual decision to list a water body is made based on assessment data. Although no comments were noted that specifically addressed "evaluative" versus monitoring data, comments showed a clear preference for the use of quantitative data that addressed specific numeric criteria.

Several organizations also commented on the need to use only quality assured data, supported by quality assurance program plans. These comments suggested that the data used should be credible and appropriate if these data are to be used to assess exceedances of WQS. It was requested that EPA provide guidance on minimum data requirements with respect to both data quality and quantity for listing and delisting.

3.2.3 Assessment Methodology

The assessment methodology is the procedure or set of rules for comparing assessment data to the WQS. Several comments identified the need to base listing decisions on adequate numbers of samples. These comments recommended that national guidance be provided for determining adequate sample size, and further recommended that waters should not be listed when adequate numbers of samples have not been collected. Comments indicated a general desire for clarity and scientific soundness in assessment methodologies. In addition, several organizations commented that only scientifically credible methods should be used in the listing process. The term "scientifically credible," however, was not defined as part of the comments. The controversy continues over what constitutes "verified and validated" models and methodology. One set of comments recommended that the methodology should be part of the continuous planning process [303(e)], rather than 303(d) listing process.

3.2.4 List Content, Format, and Submission

Commenters were, in general, opposed to the four-part list the August 1999 rule provided. The four part rule states that the 303(d) list should contain water bodies that: (1) are impaired by a specific pollutant; (2) are impaired by pollution; (3) have had a TMDL approved, but WQS have not been attained; or (4) have technology-based efforts that are expected to permit

WQS to be obtained by the next reporting period. All but two of the comments reviewed for this project indicated that states should only list water bodies that were not meeting WQS and for which TMDLs have not been established. Several comments revealed concerns about listing pollution-impaired water bodies if a pollutant-specific TMDL could not be conducted on the water body.

Most of the commenters favored a 5-year listing cycle if delisting could occur anytime within the 5-year cycle. The 5-year cycle would conform to the permitting cycle for NPDES permits and also would be compatible with the rotating basin approach used by many states. One commenter encouraged EPA to retain the 2-year cycle for at least 4 more years (two rotations) to allow states to update their listing methodologies.

3.2.5 Delisting

The 1972 Act did not specifically address delisting. EPA, however, has issued guidance to the states for delisting water bodies. The guidance notes that existing regulations require states to demonstrate good cause for not including water bodies on their lists. Good cause includes, but is not limited to, more recent and accurate data, new or updated water quality modeling or assessment, new control equipment or elimination of discharges (EPA 2000a or b). A specific process was not provided in regulations. The Final Rule stated, however, that an impaired water body must stay on the list for a given pollutant until the water body is attaining and maintaining applicable WQS for the pollutant. The commenters all indicated that the delisting process should not be any more stringent than the listing process. In addition, delisting should be able to occur on an ongoing basis, rather than on the same 2-year interval as the listing process. Guidance on the minimum requirements for delisting was requested by several commenters.

3.2.6 Other Comments

Commenters favored a listing process that provides for and encourages watershed planning alternatives to development of TMDLs. In addition, several commenters favored an alternative listing process (separate from 303(d)) that takes into account waters that (1) cannot be classified as "pollutant" impaired; (2) are expected to attain WQS; or (3) have developed TMDLs.

3.3 TMDL Science Issues Conference

The Water Environment Federation and the Association of State and Interstate Water Pollution Control Administrators held a TMDL Science Issues Conference, on March 5–7, 2001. Almost 600 individuals participated in the conference and represented a broad array of stakeholders. After 2 days of technical sessions, attendees were asked to identify the information gaps and possible action items to improve the TMDL process. Table 3-3 lists the responses recorded during the closing session.

Table 3-3. Comments Received during the Closing Session of the TMDL Science Issues Conference, 2001

TMDL Issue	Information Gaps and Proposed Action Items
WQS	1. The current WQS program is scientifically weak. Standards are not accurate, sufficient, nor scientifically valid/defensible. 2. The establishment of designated uses is flawed, and uses need to be refined. There is not enough guidance or encouragement from the federal level for refinement or subclassification of uses. The use attainability analysis process is not adequate to address the need. 3. Water quality criteria need to include realistic, defensible margins of safety. Water effects ratios for criteria are inadequate. 4. Translators for narrative standards are inadequate, and there are too many imprecise narrative criteria. 5. Antidegradation squelches tiered approaches and adaptive management, or results in too small a load reduction.
Listing	1. There is not a legitimate and working process for delisting waters. 2. There are no consistent criteria/methods for listing water bodies (especially for fish consumption advisories based on contaminants in fish tissue).
Monitoring and Data Gaps	1. Overall, there is a lack of adequate monitoring data, lack of targeted monitoring to support modeling, and inadequate ambient monitoring. 2. States and other agencies need guidelines for data collection and data management system design. 3. States are using old and inappropriate data, creating the need to better identify the quality of data that is currently being used. 4. There are not enough effective data-sharing mechanisms. 5. There are significant data gaps regarding nonpoint sources. For example, it is difficult to quantify the delivery of pollutants from nonpoint sources to the receiving water body. 6. There are not enough data on wet and dry conditions, or seasonable variations, in flow and water quality.

3.4 Congressionally Authorized Reviews

3.4.1 General Accounting Office 2000 Report

Some of the issues surrounding the TMDL program were highlighted in a General Accounting Office (GAO) report requested by the House Subcommittee on Water Resources and Environment. The results of this investigation were summarized in testimony to Congress on February 10, 2000 (GAO 2000). The primary finding of the GAO on the listing process was that the states do not have the data they need for the 303(d) listing process (GAO 2000). "Only 6 out of 50 states responding to the GAO nationwide survey indicated that they have a majority of the data needed to fully assess their waters, raising questions as to whether the states' 303(d) lists accurately reflect the extent of the pollution problems in the nation's waters" (GAO 2000). The report concluded that a critical review of the entire TMDL process was warranted, but that particular emphasis should be devoted to the actions that initiate, and terminate, the TMDL process — 303(d) listing and delisting of water bodies.

3.4.2 GAO 2002 Report

Congress also recently requested the GAO to (1) identify and assess the effects of any differences in states' approaches to identifying impaired waters; (2) determine how states ensure the quality of the data used, and (3) assess the reliability of the information in the EPA data base of impaired waters (GAO 2002). The GAO in its investigation used part of the state survey developed for this project (See Chapter 4.0).

The GAO concluded that approaches used by states to identify impaired waters varied widely. A number of factors contributed to the variability in approach among states, including different: (1) water quality standards (including designated or beneficial uses and criteria) for determining which waters are impaired; (2) types of monitoring practices used to ascertain whether these standards are exceeded; (3) procedures used to assess water quality data to make listing decisions; and (4) guidance from EPA regions for removing waters from state lists of impaired waters. This variation leads both to inconsistencies in the listing of impaired waters and to difficulties in identifying the total number of impaired waters nationwide and the total number of required TMDLs. In several cases, a common body of water, shared by two states, was listed as impaired by one state but not by the other. While there is some flexibility in listing approaches that are appropriate for different ecological and other conditions, there was no scientific basis for some of the variations.

The GAO determined that states apply a range of quality assurance procedures to ensure that data used to make impairment decisions are of sufficient quality. However, the procedures vary in their rigor. In addition, states have become increasingly vigilant about applying quality assurance procedures to the data they use from other sources.

Owing in part to the inconsistencies in states' approaches to identifying impaired waters, the information in EPA's database of impaired waters is of questionable reliability. The numbers of impaired waters cannot be compared from one state to the next, and EPA cannot reliably tally the number of TMDLs that must be completed nationwide.

3.4.3 National Research Council 2001 Report

The U.S. Congress also requested that the National Research Council (NRC) assess the scientific basis of the TMDL program, including:

Information required to identify sources of pollutant loadings and their respective contributions to water quality impairment;

Information required to allocate reductions in pollutant loadings among sources;

Availability and reliability of information for use by the states; and

Methods for obtaining this information if it is not available and reliable.

The NRC initiated its study in January 2001 and submitted the final report in June 2001 (NRC 2001). The NRC TMDL Committee originally intended to focus only on the four issues listed above. However, the Committee decided it was impossible to discuss these issues without also discussing the scientific basis for establishing WQS and the listing process that initiates the TMDL process. The delisting process was not addressed by the Committee.

The NRC TMDL Committee recommended several changes in the methods of establishing WQS and determining and listing impaired water bodies. One recommendation was that the states develop appropriate water body use designations prior to assessment and refine these use designations, if needed, prior to TMDL development. The Committee thought the "fishable-swimmable" goals of the Clean Water Act were too broad to be operational. Greater stratification of designated uses should occur at the state level because some designated uses may not be appropriate for the current and future use of the waters. Use attainability analyses, rather than TMDLs, should be conducted when the designated uses are not appropriate.

The Committee thought greater attention also needed to be given to the criterion that is selected to represent the designated use of the water body. In particular, biological criteria should be used in conjunction with physical and chemical criteria to determine if the WQS are being attained. These criteria should be defined in terms of frequency, duration, and magnitude. It was also recommended that two lists be developed: A preliminary list and the 303(d) list. The preliminary list would be for those waters where there is a lack of adequate WQS or data and analysis and additional monitoring is needed (a list of waters slated for TMDL development). Evaluated data and violation of narrative standards should not be used by themselves to add water bodies to the action list, but might be useful in placing these water bodies on the preliminary list. Finally, statistical approaches should be considered for assessing the frequency, duration, and magnitude of exceedances in defining impaired waters for listing.

3.5 FWQC/AFBF/AMSA 2002 Report

The Federal Water Quality Coalition (FWQC), the American Farm Bureau Federation (AFBF), and the Association of Metropolitan Sewerage Agencies (AMSA) sponsored preparation of a document entitled *Preparation of integrated water quality monitoring and assessment reports: Recommendations for Clean Water Act 303(d) and 305(b) methodologies and reporting* (FWQC/AFBF/AMSA, 2002). This document noted that recent listing guidance provided by EPA (EPA, 2001) was too general and did not address a number of details that states must address when revising their methodologies and developing their integrated reports (i.e. combined 303(d) and 305(b) reports per EPA's 2001guidance). The document's purpose was to "attempt to address some of those details and provide sound, science-based recommendations for States to use during the listing process." The recommendations were grouped by issue. The issues and selected recommendations are summarized as follows:

- Pre-listing evaluation of water quality standards
  - States should undertake an evaluation of their current water quality standards to determine whether the standards can be achieved and assessed.
- Collection of data and information
  - States should define quality assurance and quality control protocols that assure that data are of known and appropriate quality. States should have the ability to quickly determine the quality of different sources of data submitted in support of 305(b) assessments or 303(d) listing decisions.
- Evaluation of existing data and information

- State methodologies should provide in-depth explanations of the types, quality and quantity of data required for assessments. States should establish:
  - a hierarchy scheme for weighting acceptable sources of data,
  - requirements for data submission by stakeholder and third parties,
  - requirements for representativeness and minimum number of samples,
  - a process for addressing unclassified waters,
  - a methodology for evaluating exceedances of WQS that considers Type I and Type II errors, and
  - a weight of evidence approach.

- ♦ Development of integrated reports
  - States should apply EPA listing guidance (EPA, 2001) in developing the format of their integrated reports.
- ♦ Delisting procedures
  - States should establish an effective delisting procedure that provides for delisting between listing cycles and clearly articulate the basis for delisting.

CHAPTER 4.0

STATE REVIEWS AND SURVEYS

4.1 Rationale

Listing and delisting water bodies is fundamentally a state activity. Although EPA establishes the regulatory framework and has authority to approve and disapprove the lists, the states initiate the process through their general 305(b) and targeted 303(d) assessments. Because these activities reside with the states, the project team determined it would be useful to examine how states proceed through this listing and delisting process, what interactions they have with the EPA Regions in this effort, and what factors are considered in listing and delisting water bodies. States may have a range of answers to questions such as:

- What is the procedure for placing a water body on the 303(d) list?
- What criteria are used for listing?
- What information is evaluated for listing?
- What constitutes a WQS violation?
- Are the criteria and methods for determining impairment scientifically defensible?
- How is a water body removed from the list?
- Are the criteria the same for listing as for delisting?
- Are the listing processes similar among states?
- Are consistent review procedures used among EPA Regions?

The listing process was not expected to be consistent among states because states differ in how they establish designated uses, WQS, monitoring designs and networks, approaches for using evaluated and modeled data, and approaches for using biological information. Some of these differences were noted in the comments to EPA on both the proposed and final TMDL regulations and by the TMDL FACA Committee. These differences were also confirmed in the GAO 2002 report. (See Section 3.4.2.) To gain additional insight into the listing and delisting process for this effort, a number of states were selected and asked to participate in a survey about the TMDL process, with special emphasis on listing and delisting.

4.2 State Survey

4.2.1 Selection Criteria

Initially the project team decided to survey one pair of contiguous states in each of four regions of the country (North, East, South, and West). This also built on the geographic location and expertise of the Project Team (Limno-Tech North; TPMC-East: FTN-South; and Ross & Assoc.-West). The selected states represented a range in physiography, criteria for listing and delisting, data available for use in listing and delisting, aquatic systems, and rationale for listing and delisting.

Pairs of contiguous states were selected because contiguous states have similar physiographic or ecoregional patterns, and, therefore, might be expected to have similar water quality problems and standards. Similar WQS among states, however, are the exception more than the norm. A paired comparison between states can highlight some of the differences that should be addressed in formulating alternative listing/delisting approaches. Furthermore, a survey of contiguous states in the same EPA region could indicate if the listing/delisting process is similar for states within the same region. If contiguous states are in different EPA regions, the survey could indicate whether the regions are using different approaches in similar situations.

In addition, contiguous states may have interstate water bodies, so the survey could provide insight into how state listing approaches vary for the same water body, and how a more uniform process might be established. Finally, contiguous states can be evaluated at minimal additional costs because the Project Team had experience in working in each of these state pairs.

4.2.2 Selection of States

The final pairs of states were (1) Arkansas/Louisiana; (2) California/Oregon; (3) Michigan/Indiana; and (4) Rhode Island/Massachusetts. These particular state pairs were selected, in part, because of their willingness to participate in the study, as well as because of the experience and familiarity of the project team with those states. These state pairs were selected for other specific reasons, as well, which are outlined in the sections that follow:

4.2.2.1 Oregon and California

- Oregon has a statewide approach for establishing WQS whereas California's WQS are established using regional water quality authorities.
- Both states have lakes, rivers, and estuarine systems.
- California has one of the more stringent regulatory programs and has established regulatory regions within the state.
- Oregon is in EPA Region 10; California is in EPA Region 9. This contrast might permit an evaluation of how the review process might differ among EPA Regions.

4.2.2.2 Arkansas and Louisiana

- Arkansas uses an ecoregional approach for establishing designated uses and WQS.
- Ecoregion mineral WQS have resulted in many UAAs being conducted in Arkansas to change the designated uses and WQS. This can provide information on how the standards for listing/delisting can be changed.

- Louisiana has had multiple TMDLs conducted that are not being approved by EPA because the WQS cannot be achieved through load reductions.
- Both states are in EPA Region 6 so consistency of Regional review can be evaluated.
- Both states have reservoirs, streams, and oxbow lakes. Louisiana also has estuarine systems.

4.2.2.3 Massachusetts and Rhode Island

- Both states have inland and coastal water bodies.
- Massachusetts uses an ecoregional approach for establishing WQS, while Rhode Island has a statewide approach.
- Both states are in EPA Region 1 so consistency of regional review can be evaluated.
- Massachusetts uses probabilistic monitoring, in contrast to targeted monitoring in the other states.

4.2.2.4 Michigan and Indiana

- Both states contain rivers, lakes, and are part of the Great Lakes basin.
- Both states are in EPA Region 5 so that consistency of regional review can be evaluated.

4.2.3 Background Listing Information on Surveyed States

This section presents basic background listing information for the surveyed states. The information was obtained from EPA's TMDL Tracking System Database (EPA 1998b). The purpose is to evaluate similarities and differences between state pairs, between the selected states as a group, and among states nationwide. Given the possible inadequacies that were identified by the GAO (Section 3.4.2), the TMDL Tracking System Database may not provide a reliable basis for a detailed comparison of states. Accordingly, this presentation of background information explored general patterns of the distribution of major causes of impairment among surveyed states and all states nationally.

4.2.3.1 Number of Impaired Waters

The number of impaired waters in the surveyed states ranged from 51 (Arkansas) to 1183 (Oregon). This range is comparable to the 10th and 90th percentile ranges of impaired waters in all states. The median number of impaired waters among the surveyed states was similar to the median number for all states (310 versus 348; Table 4-1).

Table 4-1. Summary of Number of Impaired Waters in Each Surveyed State Compared to Selected Percentile Values among All States

Surveyed States	Number of Impaired Waters	National Number of Impaired Waters Percentile 10th	National Number of Impaired Waters Percentile Median	National Number of Impaired Waters Percentile 90th
Arkansas (Min)	51	112	348	962
Rhode Island	127			
Indiana	209			
Michigan	272			
Louisiana	348			
California	509			
Massachusetts	907			
Oregon (Max)	1183			
Median	310			

4.2.3.2 Major Causes of Impairment

Fifteen causes of impairment account for 90% of impairments nationwide and 92% of impairments among the states surveyed. These major causes of impairment appear on the abscissa of Figure 4-1, which compares the percent occurrence of these impairment causes in all states to their occurrence in the surveyed states. The figure shows that, in general, the top six impairments nationally are slightly underrepresented in the surveyed states whereas the next nine impairments are slightly overrepresented in the surveyed states.

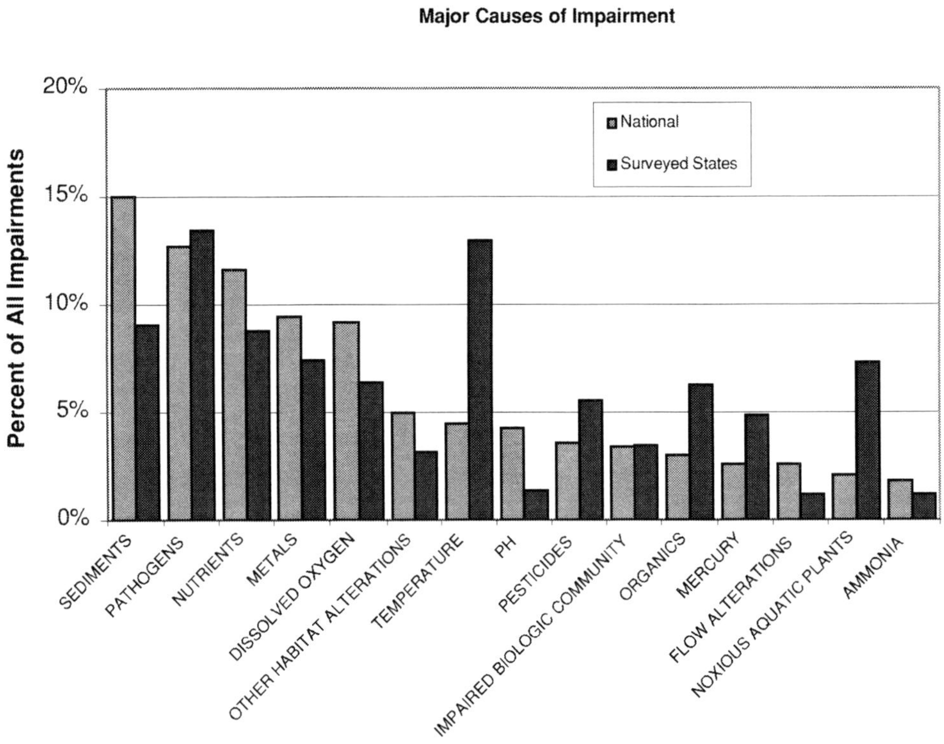

Figure 4-1. Percent Occurrence of 15 Major Causes of Impairment in All States Compared with Surveyed States

The percent occurrence of the top 15 impairments is compared between states in each state pairing in Figures 4-2 through 4-5. These figures indicate that:

- In general, states in similar geographic areas have causes of impairments that are more similar to one another than they are to states in other pairs.
- The dominant causes of impairment can vary dramatically among contiguous states (e.g., nutrients and noxious plants are significantly different between Rhode Island and Massachusetts (Figure 4-2), temperature between California and Oregon (Figure 4-3), mercury between Arkansas and Louisiana (Figure 4-4), and organics between Indiana and Michigan (Figure 4-5).

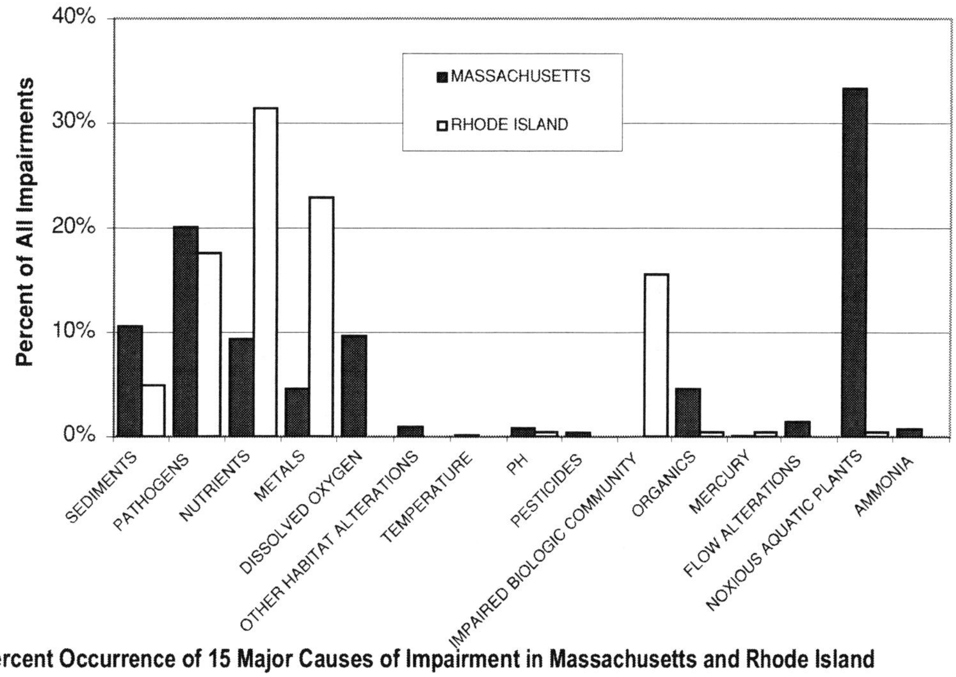

Figure 4-2. Percent Occurrence of 15 Major Causes of Impairment in Massachusetts and Rhode Island

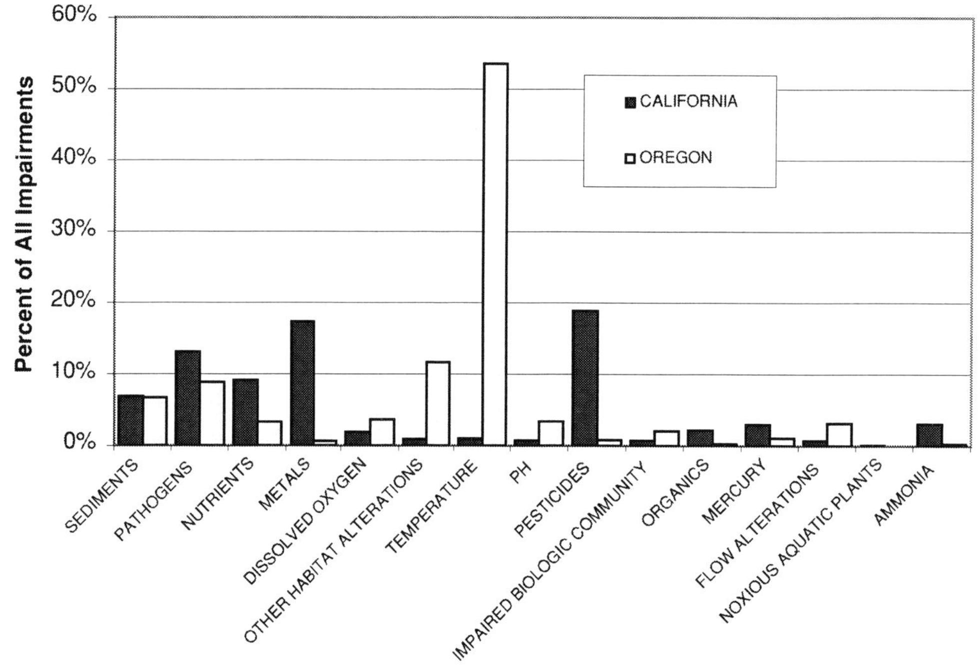

Figure 4-3. Percent Occurrence of 15 Major Causes of Impairment in California and Oregon

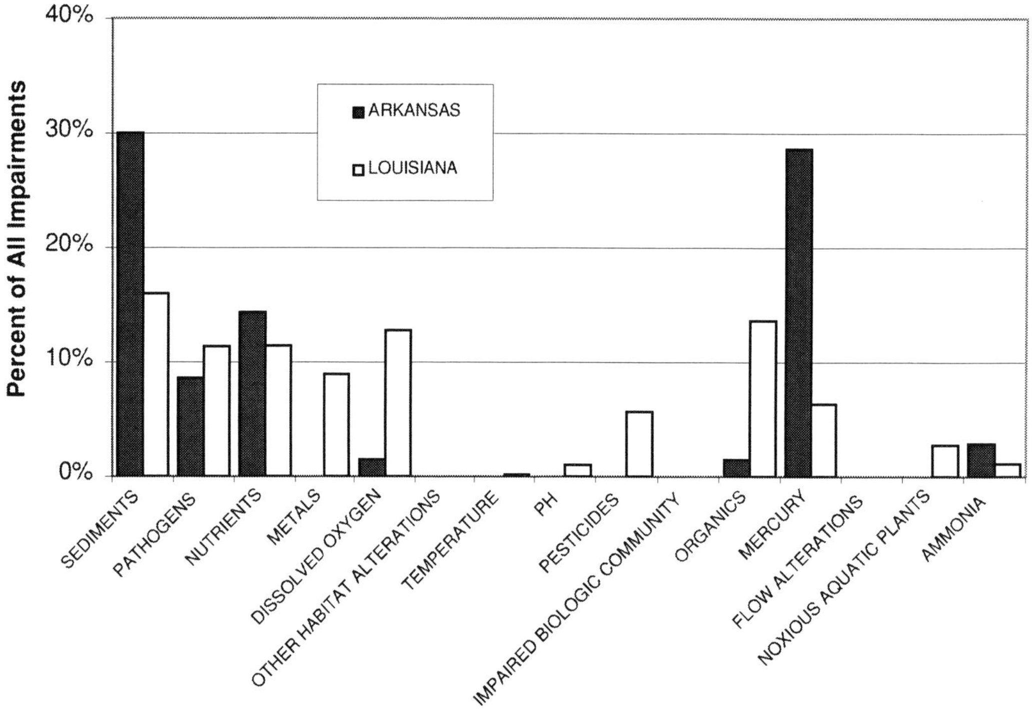

Figure 4-4. Percent Occurrence of 15 Major Causes of Impairment in Arkansas and Louisiana

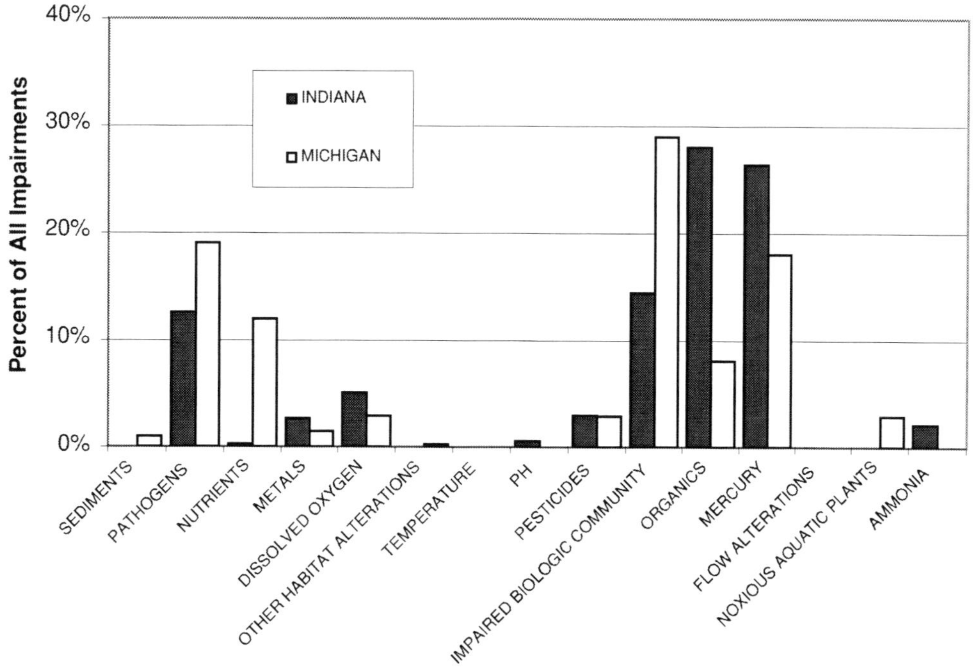

Figure 4-5. Percent Occurrence of 15 Major Causes of Impairment in Indiana and Michigan

In general, based on the number of water body impairments and causes of impairment, the surveyed states appear to provide a representative cross-section of all states nationwide.

4.2.4 *A Priori* Comparison Criteria (Identification of Listing/Delisting Issues)

Stakeholder comments (Chapter 3.0) were reviewed by the Project Team to obtain a list of common concerns among the stakeholders. The first step in the identification of issues was to develop an initial list of "hypotheses" to investigate. The intent in developing the original list of hypotheses was to write clear statements describing what was believed, at the time, to be the problems with the listing/delisting process, and then examine the evidence to support or refute each of these hypotheses. The list of hypotheses was developed based not only on stakeholder comments (Chapter 3.0), but also on the experience of the WERF investigators and Project Subcommittee (PSC) members. Once the hypotheses list was reviewed and finalized, it was converted into a list of issue questions that were included in the survey. The list of issue questions is presented in Table 4-2.

Table 4-2. Issues and Questions That Formed the Basis for the State Surveys

1.	Are WQS based on ecoregional or physiographic criteria that incorporate factors related to natural assimilative capacity of the water bodies?
2.	Do WQS incorporate both dry and wet weather conditions and criteria?
3.	Have designated uses for water bodies been demonstrated scientifically to be attainable?
4.	What are the underlying scientific bases linking designated uses to WQS (criteria)?
5.	Do narrative criteria used for listing have objective evaluative procedures and approaches for determining that designated uses are impaired and are they consistently implemented among water bodies?
6.	Is water body listing based on statistical criteria that incorporate Type I and Type II errors for different pollutants?
7.	Are evaluative procedures for assessing and listing water bodies verified and validated through scientifically designed field investigations?
8.	Is information for listing of impaired waters and WQS exceedances based on quality assured data using accepted analytical procedures?
9.	Is delisting obtained from a statistically designed monitoring network or based on verified and validated evaluative procedures?
10.	Are delisting criteria based on statistical criteria that incorporate Type I and Type II errors to ensure that WQS are being attained?

4.2.5 Pilot Survey Test

To aid in developing unambiguous survey questions to serve project needs, a pilot survey was conducted in cooperation with representatives of Mississippi and Michigan. The pilot survey was completed by state officials and a follow-up telephone interview was conducted. The purpose of the follow-up interview was to obtain further information on answers that required clarification and to receive feedback for improving the clarity of the survey. Information from the pilot survey suggested rewording some questions.

4.2.6 Conducting the Survey

The state survey forms were partially completed by Project Team members using information obtained from each state's Website. The partially completed surveys were then sent to each state 303(d) coordinator for completion and verification of the Website information.

During June 2001, personal contacts and interviews were conducted with the state 303(d) coordinators to complete the survey and confirm the process used for listing and delisting of water bodies for each of the system types — streams, lakes/reservoirs, and estuaries. The information gleaned from the surveys was summarized and synthesized to determine the various processes used by the states in listing and delisting water bodies.

4.3 Investigation and Synthesis

This section synthesizes and summarizes the state survey results in terms of this project deliverable. This survey does not represent a statistical sampling of state programs. Rather it is a targeted survey intended to identify a range of state approaches to listing issues. This section is structured based on the four primary program components related to the listing process:

1. WQS;
2. Assessment methodology;
3. Assessment data; and
4. List content, format, and submission.

Because the focus of this project was on technical issues, the questionnaire submitted to the states focused on the first three components: WQS, assessment methodology, and assessment data. The responses to the questions identified in Table 4-1 were then associated with the appropriate program component. Each issue was evaluated in light of the results from the state surveys. Whereas this chapter presents the results and findings from the survey, the next chapter presents and discusses alternative approaches to reduce uncertainty in each of these components and navigate through the listing and delisting process.

For greater ease of referencing the survey, the states' responses were entered into a spreadsheet matrix. One column of the matrix was reserved for a summary of responses from the states for each survey item. This summary matrix is provided in Appendix A. References to sections in the summary matrix in Appendix A are provided parenthetically in the following subsections of this document.

State input was solicited from 12 states, three of which declined to participate. Of the remaining nine states, only one participated in the trial survey to test the clarity of survey questions. The responses from most states were guarded. Conversations were cordial, but information was not readily forthcoming. Most of the states surveyed had just been through a 1- to 2-year settlement process because of litigation. Two of the three states that declined stated that they did not want to provide information that might impact these settlements or result in additional litigation that would affect their limited time and resources. In hindsight, this is not surprising, but it did significantly affect the information received from the states and the Project Team's ability to integrate and synthesize results and obtain consensus on needed changes to the listing/delisting process.

4.3.1 WQS (Designated Uses and Criteria)

The decision to list a water body depends on whether the water body meets WQS. Some of the comments on EPA's regulations reviewed in Section 3.2 directly addressed the issue of whether designated uses or water quality criteria are appropriate. Other reviewers of the TMDL process (e.g., NRC, attendees of the TMDL Science Issues Conference) also raised this issue. Because the determination of attainment or nonattainment with respect to WQS is the purpose of the listing process, it is crucial that the process be based on valid and appropriate WQS.

Accordingly, the survey asked state 303(d) coordinators to comment on the appropriateness, in general, of the state's designated uses and water quality criteria for particular water bodies. Survey questions were designed to obtain information on whether the states' WQS (designated uses and criteria) take into account ecological, climatologic, seasonal, and physiographic variation and the extent to which state monitoring and listing personnel see revisions of some designated uses and/or criteria as being warranted. Below is a synthesis of responses to survey questions related to WQS.

Issue Number 1: Are WQS based on ecoregional or physiographic criteria that incorporate factors related to natural assimilative capacity of the water bodies?

ISSUE SIGNIFICANCE: No states are homogenous with respect to physiography (e.g., elevation, gradient, geology) and water body types. Water quality criteria can reflect this physiographic variation, or they can be uniform statewide. Uniform statewide criteria that do not take physiographic variation into account can be overprotective for some water bodies and underprotective for others, resulting in inappropriate listing decisions for particular water bodies. In addition, some water quality criteria are dissimilar across state boundaries even though physiographic factors do not change. Water quality criteria that do not consider physiography may be more likely to be dissimilar across state boundaries.

FINDINGS (REFERENCES TO THE APPENDIX A SURVEY SUMMARY ARE PROVIDED IN PARENTHESES):

- All states surveyed had some criteria (e.g., criteria supporting human health and/or primary contact recreation uses) that were uniform for the entire state (A2.1.2).
- In some states, criteria were not uniform but were based on very coarse classifications of water bodies (e.g., trout streams vs. warm water streams; Great Lakes basin vs. other waters; oyster production vs. fresh water, A2.1.1).
- Two states surveyed based WQS on ecoregions explicitly (A2.1.0).
- In one state regional authorities set WQS. This organizational structure can result in standards and criteria that are, in practice, based on ecoregions (A2.1.0).
- In most cases, the same designated uses were assigned to all water bodies of the same general type (e.g., cold water fisheries, shellfish propagation). One state classified its "fisheries" designated use based on ecoregion (A2.3.1).
- The states surveyed did not indicate that changing WQS across state boundaries was a major concern (A2.1.4).

CONCLUSIONS: All states surveyed recognized that physiographic variation can affect designated uses and criteria, but there were significant differences among states in how WQSs were determined for water body segments. In the absence of complete WQS uniformity across states, problems in establishing the water quality status in waters that states share in common may be inevitable. The states surveyed indicated that this situation was not a serious concern. However, it should be noted that this situation is a cause for concern in other states not surveyed (GAO 2002).

Issue Number 2: Do WQS incorporate both dry and wet weather conditions and criteria?

Issue Significance: Concentrations of many water quality parameters, particularly those associated with suspended solids, change dramatically during wet weather events. Water quality exceedances may be large but also of relatively short duration. Short-term, high concentrations of some parameters observed during wet weather events may or may not impair designated uses.

Findings:

In general, states surveyed did not distinguish between wet and dry conditions in setting WQS (A2.2.0).

- Two surveyed states had wet and dry weather criteria for fecal coliform bacteria (A2.2.0).
- In one surveyed state, wet and dry conditions may be considered when assessing worst-case scenarios (A2.2.0).

Conclusions: In general, the states surveyed did not consider hydrologic variability in establishing water quality criteria.

Issue Number 3: Have designated uses for water bodies been scientifically demonstrated to be attainable?

Issue Significance: Designated uses represent the water quality management goals for a water body. If a designated use for a water body is inappropriate or unattainable, then inappropriate listing decisions may result.

Findings:

- Two surveyed states indicated that designated uses for water bodies had not been scientifically demonstrated to be attainable (A2.3.1).
- Seven states surveyed indicated that there had been some attempt (e.g., field surveys, public input) to determine if particular designated uses were appropriate (A2.3.1).
- Some states surveyed had not performed scientific analyses that were specifically intended to address the attainability of particular designated uses (A2.3.1).
- In some surveyed states, all designated uses were assigned to all water bodies in all places, while in other states surveyed, designated uses were assigned to water bodies based primarily on the maintenance of warm or cold water biota (A2.3.1).
- Some surveyed states indicated that few, if any, water bodies needed designated-use revisions (i.e., special circumstances might exist such as in transition zones between water body types). Other states surveyed indicated that some water bodies were in need of revised use designations (A2.3.6).
- Some surveyed states have used the UAA process to alter WQS. A total of about 140 UAAs have been conducted by six of the states surveyed (A2.3.5). When states mentioned applying the UAA process, they focused on site-specific criteria rather than designated uses (A2.3.5).

Conclusion: With the exception of those limited number of water body segments for which UAAs have been performed, the states surveyed have not, in general, performed scientific studies explicitly to determine if designated uses assigned to water body segments are appropriate for those water body segments.

Issue Number 4: What are the underlying scientific bases linking designated uses to WQS (criteria)?

ISSUE SIGNIFICANCE: Water quality criteria are the means for linking designated uses to actual water quality conditions. Ideally water quality criteria are closely linked to designated use attainment. Stakeholder comments (Table 3-2) indicated concern that some kinds of criteria or information (e.g., narrative, biological criteria, fish consumption advisories) should not be used for listing without supporting information. In some cases, numeric criteria do not exist for important classes of pollutants. For example, although 12.7% of all impairments nationwide are due to "nutrients" (Figure 4-1), many states do not have numeric nutrient criteria.

FINDINGS:

- EPA national water quality criteria were widely applied by all states in the survey (A2.4.1.2).
- EPA national water quality criteria were often modified based on site specific data or state-specific risk factors (e.g., fish consumption) (A2.4.1.1; A2.4.1.3).
- In some states surveyed, designated uses of a significant number of water bodies focus on cold water biota. In these cases, criteria were established to support salmonid populations (A2.4.1).
- States surveyed established water quality criteria by
    - using literature information on specific pollutant effects on warm or cold water biota or human health (A2.4.0);
    - reviewing site-specific studies (e.g., water effects ratios; A2.4.3); and
    - using a reference stream approach (collect data to establish water quality criterion values that support designated uses; A2.4.5).
- None of the states surveyed had adopted numeric biological criteria (A3.8).
- One state surveyed used biological community metrics as quasi-criteria to define aquatic life designated uses for different ecoregions (A3.8).
- Some states surveyed verbally indicated reluctance to adopt numeric biological criteria. They preferred to use biological metrics to support determinations of aquatic life use attainment.
- One state surveyed had specific plans to develop numeric biological criteria.
- All states surveyed considered existing water quality criteria protective of designated uses. However, some indicated a need for revised criteria or more criteria (A2.4.2).
- Most of the states surveyed had experience in developing site specific criteria. Site specific criteria were developed primarily for dissolved oxygen, temperature, minerals, and metals (A2.4.3).

CONCLUSIONS: Although they consider existing criteria to be protective of designated uses, the states surveyed had experience with, and were willing to develop, site-specific criteria. There was not a strong desire among the states surveyed to adopt numeric biological criteria. These states prefer at this time to use biological data as narrative criteria to use in support of listing decisions. In many of the surveyed states, water quality criteria are focused on protection of cold water fisheries.

Issue Number 5: Do narrative criteria used for listing have objective evaluative procedures and approaches for ensuring designated uses are impaired?

ISSUE SIGNIFICANCE: Stakeholder comments indicated concern over the use of narrative criteria without supporting information (Table 3-2). Narrative criteria can play an important role in the listing process. For example, according to EPA's TMDL tracking system, 20% of impairments nationwide involve "sedimentation/siltation" or "other habitat alterations" (Figure 4.1). Most of these impairments are probably based on narrative criteria because few, if any, numeric criteria exist for these kinds of impairments.

FINDINGS:

- All states had narrative criteria (A3.13).
- Seven of the states surveyed used narrative criteria to list water bodies (A3.13) while one state indicated that it did not base listing on narrative criteria.
- The surveyed states that use narrative criteria in listing appeared to avoid listing water bodies based solely on narrative criteria (A3.13.1).
- For listing purposes, the states surveyed tended to prefer that narrative criteria be supported with best professional judgment or other data such as biological assessments or information on numeric criteria (A3.13.1).

CONCLUSION: The states surveyed indicated that narrative criteria are useful for identifying problem areas but they have a clear preference for basing listing decisions on monitoring or site-specific data related to numeric criteria.

4.3.2 Assessment Methodology

Assessment methodology refers to the procedures and protocols used to compare monitoring or evaluated data to WQS.

Issue Number 6: Is water body listing based on statistical criteria that incorporate Type I and Type II errors for different pollutants?

ISSUE SIGNIFICANCE: The comparison of monitoring data to a criterion fundamentally is a hypothesis test. A fundamental feature of hypothesis tests is that the conclusions of the test may err, that is be "inaccurate", in one of two ways[1]. In the context of listing, a Type I error occurs when an unimpaired water body is classified as impaired. A Type II error occurs when an impaired water body is classified as unimpaired. All comparisons, whether they are statistically based or not, have Type I and Type II errors associated with them. A methodology that does not incorporate these errors will list unimpaired waters and fail to list impaired waters at unknown rates. A statistically rigorous procedure allows the tradeoffs between Type I and II errors to be considered in defining the rules for classifying water bodies as impaired or not impaired. A statistically rigorous assessment methodology begins with an explicit choice of which type of error to fix and what fixed error rate is appropriate. This choice is a policy decision. This choice has a direct bearing on the number of TMDL studies conducted.

The state survey did not specifically inquire about whether the state formally considered statistical error rates in assessing water bodies. Rather, the survey inquired about the "standard of

[1] Statistical tests have two types of error associated with them: Type I and Type II error. Type I error occurs when the null hypothesis (the site is not impaired) is rejected, and a water body is determined to be impaired when it is not. The Type II error occurs when the null hypothesis is not rejected even though the water body is impaired. For a given comparison, Type I and Type II errors are complements. Therefore, reducing one type of error increases the other type of error.w

proof" that states prefer in making assessments. Specifically the survey asked if the states favored a policy that requires relatively strong evidence of impairment to list the water body or applies a protective or precautionary principle and lists a water body even if there is only modest indication of impairment. These choices correspond to protection of Type I and Type II error, respectively.

FINDINGS:

- Although the states surveyed had explicit procedures for comparing monitoring data to numeric criteria (typically based on percent exceedances and/or maximum or minimum values) none of the states' assessment methodologies were based on formal statistical procedures (A3.5).
- None of the states surveyed explicitly considered error types in their assessment methodology (A3.4).
- Three of the states surveyed indicated that they used or favored application of the precautionary principle (this approach would correspond to protection of Type II error) for assessment, but two of those indicated that they did not plan to use that approach in future listings (A3.4).
- Of the remaining five states, three answered that they favored a policy that required a relatively high degree of certainty (or at least documentation collected in accordance with a quality assurance plan) that a water body is impaired before it is listed (A3.4).
- One state indicated that they apply both approaches, and a single answer could not be provided (A3.4).

CONCLUSIONS: Although any particular listing decision is fundamentally a hypothesis test, none of the states surveyed used a formal statistical approach to compare monitoring data to numeric criteria. Only two states surveyed indicated that they favored application of a protective principle (corresponding to protection of Type II error) for classifying water bodies as impaired or not impaired.

Issue Number 7: Are evaluative procedures for assessing and listing water bodies verified and validated through scientifically designed field investigations?

ISSUE SIGNIFICANCE: Evaluative procedures rely on evaluated data, which consist of "...monitored data that are more than 5 years old, as well as other types of information such as land use data, predictive models, and other less-precise indicators of water quality" (GAO, 2002). Stakeholder comments indicated concern over listing decisions that are made without quantitative evidence of impairment (Table 3.2). Evaluative procedures are another test to help ensure that all potentially impaired waters are scrutinized by the listing process.

FINDINGS:

- Evaluative procedures were widely used in listing decisions by four of the eight states surveyed (A3.11, A3.11.1, A3.11.2, A3.11.3, A3.11.4).
- One of these four states indicated that evaluative procedures would not be used in the future.
- Although some of the states surveyed indicated that evaluative procedures had been verified through special studies, none of the states surveyed conducted corroborative data collection to verify the procedures.
- The use of evaluative assessments can have a significant effect on the end results of the listing process depending on the "standard of proof" that the evaluation requires. For example, 78% of the state of Mississippi's 303(d) listings (562 of 721 listings) resulted from evaluative assessments that allowed only very modest indications of impairment to result in listings.

CONCLUSIONS: Although evaluative procedures were widely used by the states surveyed, the procedures tended to be informal and were not corroborated or verified through data collection.

4.3.3 Assessment Data and Monitoring Systems

Questions were developed for the survey to obtain information describing how data are collected by states and how quality assurance/quality control issues are addressed.

Issue Number 8: Is information for identifying impaired waters and WQS exceedances based on quality assured data using accepted analytical procedures?

ISSUE SIGNIFICANCE: Assessing WQS exceedances requires data that are comparable to applicable criteria and are of known quality. Adequate QA/QC procedures are necessary to ensure that data are collected according to methodologies that are consistent with the needs of the listing process.

FINDINGS:

- All states surveyed indicated that, wherever possible, they used data that have passed through some QA review (A4.1, 4.2, 4.3, 4.4, 4.5).
- All states surveyed had QA/QC programs in place for the data collected as part of the state monitoring programs.
- All states surveyed considered the data used in making listing decisions to be reliable, especially if data were collected relatively recently (less than 5–10 years old) (A4.4).

CONCLUSIONS: All states surveyed recognized the need for QA/QC programs to control data quality, but the rigor of the QA process varied among states.

4.3.4 Delisting

Unlike the listing process, neither EPA nor most states have issued clear procedures or guidance for delisting of water bodies. Questions were included in the state survey to obtain information describing how the states are presently delisting water bodies. Issues 9 and 10 are considered together in the following text.

Issue Number 9: Is delisting obtained from a statistically designed monitoring network or based on verified and validated evaluative procedures?

Issue Number 10: Are delisting criteria based on statistical criteria that incorporate Type I and Type II errors ensuring that WQS are being attained?

ISSUE SIGNIFICANCE: *The same issues that pertain to the use of quantitative and evaluative data, and Type I and Type II errors for listing also apply to delisting (See Issues 6 and 7).*

FINDINGS:

- Three of the states surveyed had not delisted any water bodies (A5.1). The number of delisted water bodies in four of the remaining states surveyed ranged from 3 to >100 (3, 42, 68, >100). One state reported delisting of 476 miles of previously impaired waters (A5.1).
- Delisting was not based on data collected from statistically designed monitoring networks or based on verified and validated evaluative procedures.
- None of the states surveyed used evaluative procedures for delisting.
- None of the states surveyed considered statistical criteria that incorporate Type I and Type II errors for delisting.
- Delisting approaches in the states surveyed involved mainly obtaining new or additional data, revising WQS (mainly criteria), and EPA approval of a TMDL (A5.3).
- One state surveyed had the following as part of its listing guidelines (1998 Clean Water Act (CWA) Section 303(d) Listing Guidelines for California, August 11, 1997, prepared by ad hoc workgroup of staff from the state and regional water boards, and EPA Region 9):

> "Water bodies may be delisted for specific pollutants or stressors if any one of these factors is met:
>
> (1) Objectives are revised (for example, Site Specific Objectives), and the exceedance is thereby eliminated.
>
> (2) A beneficial use is redesignated after US EPA approval of a Use Attainability Analysis, and the non-support issue is thereby eliminated.
>
> (3) Faulty data led to the initial listing. Faulty data include, but are not limited to, typographical errors, improper quality assurance/quality control (QA/QC) procedures, or Toxic Substances Monitoring/State Mussel Watch EDLs which are not confirmed by risk assessment for human consumption.
>
> (4) It has been documented that the objectives are being met and beneficial uses are not impaired based upon "Monitored Assessment" criteria.
>
> (5) A TMDL has been approved by the US EPA.
>
> (6) There are control measures in place which will result in protection of beneficial uses. Control measures include permits, clean up and abatement orders, and watershed management plans which are enforceable and include a time schedule."

CONCLUSIONS: The states surveyed varied widely in their experience with delisting. In most of the surveyed states, delisting typically involved reevaluating the listing decision in light of new data or a revised WQS, or completion of a TMDL. As with listing (See Issue 6), the delisting approaches were not based on formal statistical procedures.

4.3.5 States' Suggestions for Improvements

In the survey, states were asked how they would improve certain aspects of the listing/delisting process. Indications that some aspects of WQS, particularly criteria, were in need of revision have been summarized above. States indicated ways to improve other aspects of the process as follows.

Question Number 1: How should states develop water quality criteria that are protective of designated uses?

FINDINGS:

- Five of the eight states provided feedback on this issue (A2.4.5). Responses included the following:
    - Adopt a site-specific approach;
    - Measure existing conditions that support existing designated uses;
    - Assemble expert panel; and
    - Allow criteria development to be part of the TMDL process.

CONCLUSIONS: Some of the states surveyed recognized a need for water quality criteria that are more site specific.

Question Number 2: How could the reliability of the data be improved?

FINDINGS:

- States surveyed indicated that reliability of recent data was adequate.
- More data are needed (i.e., more monitoring stations, greater sampling frequency; A 4.5).

CONCLUSIONS: States surveyed believe they have the expertise and experience necessary to generate reliable data. In general, however, they believe more data are needed to make the listing process more reliable. That is, while the quality of monitoring data is considered adequate, improvement in the quantity of data is needed.

Question Number 3: How could the listing process be improved?

FINDINGS:

- The surveyed states' suggestions for improving the listing process included (A8.3.1):
    - base criteria on high-quality data, and
    - base listing decisions only on high-quality data so that only truly impaired waters are listed.
- There was a general desire among the states surveyed to see a flexible listing process that would allow states to incorporate documented best professional judgment, weight of evidence, and site-specific knowledge into the process. This would allow states to focus resources on impaired waters.
- Some states surveyed indicated that EPA should provide guidance on technical issues (e.g., criteria development, WQS revisions, design of monitoring programs, assessment methodology) while allowing states flexibility in application and implementation.

CONCLUSIONS: The states surveyed do not wish to make listing decisions in the absence of adequate data. These states desire a listing process that is flexible, incorporates state knowledge of water quality problems and issues, and is based on technical guidance provided by EPA.

Question Number 4: How could the delisting process be improved?

FINDINGS:

- The states surveyed indicated a desire for
  - EPA guidance for delisting water bodies other than completion of a TMDL; and
  - A requirement that data or information for delisting equivalent to that required for listing.

CONCLUSIONS: The states surveyed desire EPA guidance on delisting based on data, and that data requirements for delisting be the same as those for listing.

4.4 Summary

The state survey conclusions are summarized in Tables 4-3 and 4-4. The survey results indicate a range in approaches that states use to address issues such as the adequacy of WQS, linking designated uses to criteria, hydrologic and seasonal variability, statistical comparisons, and use of evaluative data. The range of approaches to these issues indicates that there is technical and scientific uncertainty underlying the listing process. In general, the states surveyed acknowledge the listing/delisting process could be improved and are willing to provide suggestions for improvement. The next chapter presents a series of technical approaches and alternatives that have been used or might be used to reduce uncertainty, and better navigate the listing and delisting process.

Table 4-3. Summary Conclusions from the State Survey

Issue 1:	*Are WQS based on ecoregional or physiographic criteria that incorporate factors related to natural assimilative capacity of the water bodies?*
	• States surveyed recognize that physiography influences designated uses and criteria, but there are significant differences among states in how this variation was considered in WQS.
Issue 2:	*Do WQS incorporate both dry and wet weather conditions and criteria?*
	• States surveyed generally do not consider hydrologic variability in establishing water quality criteria.
Issue 3:	*Have designated uses for water bodies been scientifically demonstrated to be attainable?*
	• With the exception of a relatively limited number of UAAs, states surveyed have not explicitly determined that designated uses are attainable.
Issue 4:	*What are the underlying scientific bases linking designated uses to WQS (criteria)?*
	• All states surveyed consider existing criteria protective of designated uses but the basis for criteria varies from site specific determinations to adopting EPA national criteria.
Issue 5:	*Do narrative criteria used for listing have objective evaluative procedures and approaches for ensuring designated uses are impaired?*
	• Narrative criteria are considered useful for problem identification, but listing decisions should be made based on monitoring or site specific data.
Issue 6:	*Is water body listing based on statistical criteria that incorporate Type I and Type II errors for different pollutants?*
	• No states surveyed use a formal statistical approach to compare data with criteria.
Issue 7:	*Are evaluative procedures for assessing and listing water bodies verified and validated through scientifically designed field investigations?*
	• Evaluation procedures generally have been tested only through special studies or informally.
Issue 8:	*Is information for identifying impaired waters and WQS exceedances based on quality assured data using accepted analytical procedures?*
	• All states surveyed have formal QA/QC programs for monitoring data.
Issues 9 & 10:	*Is delisting obtained from a statistically designed monitoring network or based on verified and validated evaluative procedures? Are delisting criteria based on statistical criteria that incorporate Type I and Type II errors ensuring that WQS are being attained?*
	• None of the states surveyed use formal statistical approaches for delisting.

Table 4-4. Summary of States' Suggestions from the State Survey

Question 1:	*How should states develop water quality criteria that are protective of designated uses?* • States surveyed recognize the need for site-specific criteria and suggest: - Site-specific approaches, - Measure existing conditions that support existing designated uses, - Expert panel, and - Criteria development as part of TMDL process.
Question 2:	*How could the reliability of the data be improved?* • In general, the surveyed states believed the listing process would be more reliable with more data.
Question 3:	*How could the listing process be improved?* • Base listing on high-quality data. • Establish a flexible process. • EPA issue technical guidance.
Question 4:	*How could the delisting process be improved?* • EPA issue technical guidance for delisting based on data.

CHAPTER 5.0

NAVIGATING THE PROCESS: ALTERNATIVE APPROACHES FOR THE LISTING AND DELISTING PROCESS

Conceptually, the listing and delisting process is straightforward:

1. Identify the appropriate WQS for the water body;
2. Obtain corresponding water quality data for that WQS; and
3. Compare data to the WQS.

Then either the water body is listed if the data indicate the WQS is not being attained, or the water body is delisted if the water body is listed and the data indicate the WQS is being attained.

As has been described throughout this report, however, various issues have arisen, and multiple approaches have been used for navigating through the listing and delisting process. Although the Clean Water Act forms the foundation for protecting and maintaining the quality of waters in the United States, different states function under different water quality policies, and legislative and regulatory mandates.

Each state and territory is required to conduct a triennial review of WQS to ensure the designated uses and associated water quality criteria are protecting its waters. Although this process is independent of the listing/delisting and TMDL processes, it is the foundation of both processes. Compliance with WQS is the sole determinant for placement on the 303(d) list and subsequent TMDL development. WQS are the critical basis for the listing process because the ultimate designation of impairment is based on comparison of ambient monitoring or evaluated data with the WQS. The NRC Committee stated that the foundation of the water quality management program was based on the setting of appropriate WQS (NRC 2001). It also indicated that the generic “fishable/swimmable” goals of the Clean Water Act represented the beginning, not the end, of appropriate use designation.

In addition there is scientific and technical uncertainty imbedded in each element of the listing/delisting process. For example, the evaluation of state approaches revealed that there is wide variation in the degree of scientific rigor applied to assigning designated uses and associated criteria to water bodies. Although all states have QA/QC programs in place, there is wide variation in the rigor of QA/QC review among programs. Statistical uncertainty is very rarely incorporated in state approaches. As a result, many attainment decisions are affected by factors related to the appropriateness of the WQS, data quality, and statistical uncertainty. Consequently, the accuracy or appropriateness of any given decision to classify a water body as impaired or unimpaired can be uncertain. Therefore, alternatives and options to reduce uncertainty in WQS are considered critical not only to this report, but to the overall listing/delisting process.

The alternatives and options presented in this chapter are based on approaches suggested by the surveyed states, currently used by these and other states, and/or based on published research. They are intended to assist water quality professionals in navigating through the listing/delisting process and reducing the uncertainty in classifying water bodies as attaining or not attaining WQS. This reduction in uncertainty, however, is not without costs in resources and personnel. Reducing uncertainty through additional site investigations, data collection, and more sophisticated analysis methods all require additional time and money. State and other water quality professionals must weigh the acceptable risk of misclassification against the costs to reduce that risk.

This chapter provides:

- A minimum approach for all states to consider as they navigate the listing/delisting process;
- Options for reducing uncertainty in WQS, water quality data, comparative methodology, and list content and format; and
- A comprehensive approach that, perhaps not achievable, illustrates what might be required to fully navigate the listing/delisting process.

5.1 Minimum Approach

Based on the review of comments (Chapter 3.0) and the state surveys (Chapter 4.0), the project team identified fundamental issues that must be addressed to assure a scientifically and technically defensible listing and delisting process. Consistent with the approach used in this study, this minimum approach is organized around the four basic elements of the listing process.

5.1.1 Water Quality Standards

At a minimum, WQS should be critically reviewed, including both revisions since the last assessment and revisions proposed for the next triennial review or legislative session. The critical review should focus on the appropriateness of both designated uses and supporting criteria.

The branch or section of state government responsible for the triennial WQS review is typically not the branch or section responsible for the listing/delisting assessment process. Different states have different approaches for adopting proposed changes to the WQS and the WQS are not necessarily adopted only following the triennial review. A critical WQS review should help to ensure that the WQS appropriate for the specific water body segment or assessment unit are compared with the site-specific water quality data. Even though the proposed

WQS revisions cannot be used until formally approved, water bodies attaining or not attaining WQS following formal revision might be cited in a different category on the 303(d) list (e.g. category 4b of 2002 EPA Listing Guidance). Regardless of the individual state procedures, however, based on the Alaska rule EPA must formally approve the WQS changes before they become effective and can be used to determine impairment.

5.1.2 Data Acquisition

> *At a minimum, QA/QC requirements should be established for all data used as part of the listing and delisting process.*

All states have QA/QC requirements for data collected as part of their monitoring programs, and most states have QA/QC requirements for extant data proposed for use in the 303(d) process. Wyoming and Florida have passed state data integrity laws with specified QA/QC requirements for any data used as part of the 303(d) process. Quality assurance project plans document how QA and QC are applied to environmental data operations to assure that results are of the type and quality needed and expected and are of known quality (EPA 2001a, EPA 1994a). The state survey, subsequent state contacts, and the GAO (GAO 2002) indicated there are different interpretations among EPA Regions on what constitutes "all readily available data." If QA/QC information is not provided with the data, its quality cannot be determined and its status as "readily available data" may be uncertain. The QA/QC requirements for all data used in the assessment of impairment should define data quality objectives consistent with the intended use of the data. These objectives can form the basis for accepting or rejecting data sets for use in water quality assessments.

5.1.3 Assessment Methodology

> *At a minimum, each state should explicitly indicate its preference for controlling Type I or Type II errors.*

A Type I error occurs when a water body that is fully supporting a WQS is classified as partially or not supporting (TNRCC 2002). A Type II error occurs when a water body that is partially or not supporting a WQS is classified as fully supporting (TNRCC 2002). Texas, Florida, and Washington explicitly state the permissible Type I error rate to be used in comparing water quality data with WQS.

> *At a minimum, each state should specify the confidence level (error rate) desired (permitted) in the tests comparing the data with the WQS.*

If the error rate cannot be specified or is unknown, there is no way of estimating the uncertainty in the listing/delisting process. That is, the probability of incorrectly classifying water bodies will be unknown. Florida, Mississippi, Texas, and Washington specify the Type I error rate for determining criteria exceedances.

Specifying the error protection (Type I or Type II) and error rate provides a basis for addressing the issue of data quantity. Once error protection and error rates are specified, the sample size needed to achieve these values can be determined.

5.1.4 List Content and Format

At a minimum, states should adopt the List Content and Format categories in the EPA 2002 Integrated Water Quality Monitoring and Assessment Report Guidance (EPA 2001b).

EPA issued guidance to the states on November 19, 2001, for development of the 2002 303(d) lists. The states were subsequently given the option of using this listing guidance, their current listing approaches, or a hybrid approach for the 2002 list. If EPA adopts this guidance for future assessments, the states can gain experience in categorizing their waters using the EPA method by proceeding through the guidance for the 2002 list. As importantly, however, this permits the states to evaluate the proposed guidance, identify areas of uncertainty or weakness and provide recommendations to revise the guidance and help reduce the uncertainty of classification. A report sponsored by FWQ/AFBF/AMSA (2002) indicated many states are using a hybrid approach.

5.1.5 Minimum Approach: Summary

In summary, the suggested minimum for navigating through the listing/delisting process to reduce uncertainty is:

- ♦ Critically review WQS for appropriateness, including revisions since the last assessment as well as revisions proposed for the next triennial review or legislative session.
- ♦ Establish QA/QC requirements for all data used as part of the listing/delisting process.
- ♦ Establish a statistically based comparison methodology that explicitly considers Type I versus Type II errors.
- ♦ Specify the confidence level (error rate) desired in the tests comparing the data with the WQS.
- ♦ Adopt the List Content and Format categories in the EPA Listing Guidance.

5.2 Alternatives and Options for Reducing Uncertainty

As indicated throughout the above discussion, states are already using some of these approaches to reduce the uncertainty in the listing and delisting process. Other options or alternatives being used by the states or supported through additional research are described in this section, along with selected advantages and disadvantages of each option (Tables 5-1 through 5-4, Figure 5-1).

5.2.1 Water Quality Standards

WQS consist of designated uses, water quality criteria to protect these designated uses, and an antidegradation policy. Uncertainty associated with WQS may arise when inappropriate WQS are established for a water body. WQS may be inappropriate as a result of inappropriate designation of water body uses or establishment of inappropriate water quality criteria to protect these uses.

As indicated by the GAO, many designated uses were established during the early 1970's over a 180 day period to meet the requirements of the Clean Water Act (GAO 2002). The general goals of the Act, which were adopted by the states, were for fishable and swimmable uses for U.S. waters. Since then, many states have refined designated uses for their water bodies, but these general goals still persist for many, if not all, water bodies. For example, all of Virginia waters are designated for swimming uses even though some of the waters are too shallow for swimming (GAO 2002). Many states designate all freshwaters as drinking water supply even though some water bodies cannot provide drinking water supply. The states surveyed for this project indicated that, in general, the designated uses for their water bodies were appropriate, but also indicated that designated uses needed to be revised for some water bodies. (See Section 4.3.1.) Water bodies with designated uses that require revision can be inappropriately included on the 303(d) list.

Table 5-1. WQS Options

1. Base WQS on Physiographic Province/Ecoregion or Watershed Type (AR, OH, LA, RI)	
ADVANTAGES	**DISADVANTAGES**
• Incorporates natural environmental heterogeneity in designated use and water quality criteria • Applicable to many water bodies in the province/ecoregion	• Requires establishment of regional reference conditions • Does not eliminate potential need for site-specific uses/criteria
2. Establish Site-specific Criteria (EPA 1997, 40 CFR §131.10, KS, MT, ID, IN)	
ADVANTAGES	**DISADVANTAGES**
• Criteria appropriate for specific water body • Significant reduction in uncertainty • Can be used on an as-needed basis • Incorporates natural background	• Time and resource intensive if more than a few sites • Cannot be done for all state waters
3. Establish Seasonal or Wet Weather WQS (LA, OH)	
ADVANTAGES	**DISADVANTAGES**
• Links seasonal variability in use to criteria • Incorporates seasonal constituent variability in criteria • Reduces over- or underprotection of use and resources • Accounts for hydrologic variability in constituents • Allows exceedances due to natural flow regime • Can incorporate natural background in criteria	• Requires seasonal sampling • Data must be partitioned by seasons and/or flow • Can increase data requirements for assessment • Requires establishment of appropriate uses relative to seasonal and hydrologic variability • Time and labor-intensive to establish • Requires establishment of background
4. Establish Tiered WQS (OH, NY)	
ADVANTAGES	**DISADVANTAGES**
• Incorporates habitat variability in uses and criteria • Provides flexibility in uses • Incorporates background consideration	• Requires habitat assessments for multiple water bodies • Requires identification of appropriate use risk-categories • Requires UAA for some uses

Table 5-1. WQS Options (Continued)

5a. Biological Criteria — Use as sole basis for determining impairment (MS, OH)	
ADVANTAGES	**DISADVANTAGES**
• Direct measure/indicator for assessing aquatic life use • Integrates chronic and episodic exposures • Eliminates need for weight-of-evidence approach • Can be implemented either as narrative or numeric criteria	• Requires regional reference frame • Does not distinguish between pollution and pollutant impairment • Requires biological expertise to develop criteria and interpret data • Ignores other types of data (independent applicability not considered) • Eliminates weight of evidence approach
5b. Biological Criteria – Use in support of decisions based on physical/chemical criteria (AR, IN, LA)	
ADVANTAGES	**DISADVANTAGES**
• Corroborate or refute estimates of aquatic life non-attainment based on nonbiological data • Contributes to weight of evidence approach in assessing attainment/ nonattainment of WQS • Consistent with independent applicability principle	• Requires weighting factors or other decision-making framework to resolve conflicting data sets.
6a. Use Numeric Criteria Only	
ADVANTAGES	**DISADVANTAGES**
• Provides objective basis for assessing water quality data • Reduces ambiguity in water quality assessments	• Eliminates potential for listing waters impaired by pollutants without numeric criteria • Reduces citizen involvement
6b. Use Narrative Criteria with Translator	
ADVANTAGES	**DISADVANTAGES**
• Preserves flexibility in use of professional judgment • Links objective criteria with subjective observations • Provides means for implementing biocriteria without establishing numeric biological index criteria	• Some narrative criteria may be difficult to link to translator
7. Apply UAA Process (KS, 40 CFR §131.10 (g))	
ADVANTAGES	**DISADVANTAGES**
• Permits removing designated uses that are not existing uses • Permits changing uses before 303(d) listed • Third party initiated and conducted studies • Identifies criteria that are still protective of uses	• Time and resource intensive
8. WQS Based on Magnitude, Frequency, Duration	
ADVANTAGES	**DISADVANTAGES**
• Incorporate seasonal and temporal variability in criteria • Accommodates exceedances due to natural flow regime	• Procedures not fully ready to establish criteria • Research stage • Greater data and analysis requirements

Table 5-2. Data Acquisition Options

Advantages	Disadvantages
1. Sampling Design (MD, VA, WV)	
• Design tailored to goals, objectives, and resources • Objective basis for site selection • Flexibility in design	• Probabilistic site not applicable for 303(d), only 305(b) • Site access can be difficult • Can increase time, resource requirements
2. Response Design (OH, MS)	
• Ensures criteria constituents monitored • Captures seasonal, temporal variability in data • Incorporates biology, physiology, and chemistry	• Can increase time, resource requirements • May not permit fixed schedule sampling
3. Establish Quality Assurance Requirements Internally	
• Provides data of known quality • Permits data from multiple agencies/ organizations to be assessed	• Can eliminate some readily-available data from assessment • May increase analysis and assessment time and resources
4. Establish QA/QC Requirements through Legislation	
• Legally mandated QA/QC requirements for all assessment data • Rejection criteria clear • All data of known quality	• Loss of flexibility for including data • Requires legislation process for implementation or change
5. Eliminate Listing Based on Evaluated Information	
• Eliminates subjective opinion or observation • Requires recent monitored data for assessments • Listing based only on actual site data	• Reduces coverage of waterbodies • Can reduce public participation • May conflict with EPA guidance • Eliminates subjective opinion or observation
6. Include Third Party Data Collection (GA, UT)	
• Greater coverage of monitored waters • Increased data/ information for assessment • Public participation in the assessment process	• Requires strict QA/QC protocols • Poor public image if data rejected • Should have third-party overview for field/lab procedures

Table 5-3. Comparative Methodology Options

1. Error Rate Specifications (FL, TX, WA)	
ADVANTAGES	**DISADVANTAGES**
• Type I, Type II errors known • Misclassification probability can be estimated	• No scientific basis for error rate selection
2. Statistical Tests (FL, TX)	
ADVANTAGES	**DISADVANTAGES**
• Statistical tests can be based on data quantity and distribution • Misclassification likelihood can be estimated • Hypothesis testing procedures are well established	• Knowledge of statistics required • Tests can be sensitive to data distributions
3. Minimum Sample Size/Water Body (AR, FL, MS, OH, TX)	
ADVANTAGES	**DISADVANTAGES**
• Control Type II errors through sample size • Ensure adequate data set for assessment • Objective criteria for including/excluding data for assessment	• Additional sample requirements for many water bodies • Additional time and resource requirements
4. Use Verified Evaluative Procedures (LA, MS)	
ADVANTAGES	**DISADVANTAGES**
• Only verified procedures used for assessment • Expanded coverage for assessed waters • Uncertainty estimates associated with assessed water bodies	• Waters assessed with no monitoring data • Potential listing without site water quality data

Table 5-4. List Content and Format Options

1. Follow 19 November EPA Guidance on 2002 303(d)	
ADVANTAGES	**DISADVANTAGES**
• EPA-issued guidance: Optional for 2002, but to be used in out years • Provides prototype for testing and possible revision prior to 2004 List • Flexibility and options for assessment • Delist following completion of TMDL or reasonable expectation of attainment • Provides categories for water bodies with insufficient information	• Likely to be revised before 2004 so additional effort may not be justified • Guidance is unclear for some categories
2. Planning/Verified Lists (FL)	
ADVANTAGES	**DISADVANTAGES**
• Flexibility and options for assessment • Only verified, impaired water bodies are listed • Recommended by NRC (2001)	• Requires additional step in assessment process • Can require additional follow-up monitoring • Delist only after attaining WQS
3a. Delisting — Specified Delisting Criteria (FL)	
ADVANTAGES	**DISADVANTAGES**
• Explicit criteria for delisting water bodies • Eliminates ambiguity for delisting	• Lack of flexibility for delisting • Can be difficult to revise • Delisting can be more rigorous than listing
3b. Delisting — Phased TMDL	
ADVANTAGES	**DISADVANTAGES**
• Permits delisting without attainment based on reasonable expectation	• Requires additional monitoring and reevaluation
3c. Delisting — Adaptive Management	
ADVANTAGES	**DISADVANTAGES**
• Acknowledges attainment is long-term process • Delisting following completion of TMDL	• Permissibility under TMDL rule uncertain • Requires continued monitoring

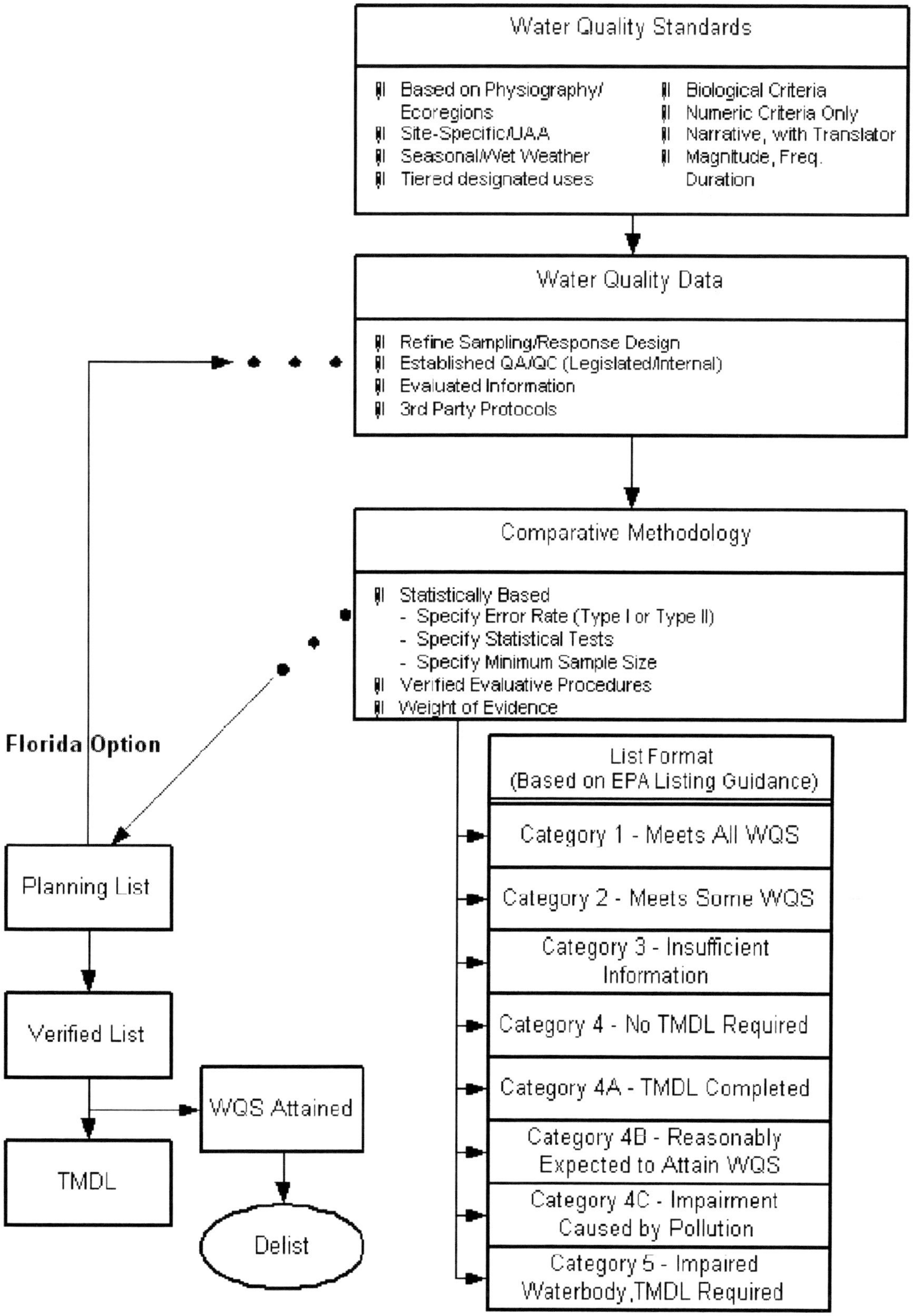

Figure 5-1. Four Major Elements of the Listing/Delisting Process and Options or Alternatives for Reducing Uncertainty and Navigating through the Process (after GAO 2002)

Uncertainty also arises from the water quality criteria linked to designated uses. For example, a dissolved oxygen (DO) criterion of 5 mg/L or higher might be established to protect aquatic life use in streams. Yet, in slow, sluggish water bodies, natural DO concentrations might only attain concentrations of 3 or 4 mg/L with a healthy aquatic community that has evolved and adapted to these conditions. Based on a DO water quality criterion of 5 mg/L, this water body would likely be included on the 303(d) list, even though it is not impaired for aquatic life use.

Refining designated uses and water quality criteria can reduce the likelihood of inappropriate listings. Nevertheless, it is unrealistic to expect that designated uses and water quality criteria will be established through site-specific investigations for all waters of the state. Several alternatives that states have used to establish appropriate WQS for water bodies and reduce the uncertainty include (Table 5-1, Figure 5-1):

- Establish WQS based on physiography, ecoregions, or water body type;
- Use site-specific WQS for selected water bodies;
- Establish seasonal or wet weather WQS;
- Use tiered WQS;
- Establish WQS based on magnitude, frequency, and duration;
- Use only numeric criteria or establish translators for narrative criteria; and
- Conduct use attainability analyses (UAAs).

Following is a discussion of each alternative listed above.

5.2.1.1 Establish WQS Based on Physiography, Ecoregion, or Water Body Type

Statewide uniformity may be appropriate for some criteria (e.g., fecal coliform concentrations for primary contact recreation; maximum temperatures for trout fisheries) or for small states (e.g., Rhode Island). In general, however, state landscapes are heterogeneous, and this physiographic heterogeneity is reflected in the physical, chemical, biological, and habitat characteristics of the water bodies (ADPCE 1987; Yoder and Rankin 1995a, 1995b). These water body characteristics affect the designated uses and associated water quality criteria that are selected to protect these designated uses. The states in the survey varied in the degree to which their WQS considered physiographic variation across the state. Ecoregion-based approaches have been used by some states to incorporate physiographic characteristics into WQS (ADPCE 1987; Hughes 1995; Hughes and Larsen 1988; Hughes et al. 1994; Omernik 1995; Omernik and Griffith 1991; Yoder and Rankin 1995a). Uniform state criteria that do not consider the physiographic characteristics of water bodies can contribute to inappropriate listing of water bodies. If natural background concentrations or values exceed criteria, it will not be possible to attain WQS, regardless of reductions in point and nonpoint source pollutant loads.

An example of how physiography-based criteria can affect the outcome of the listing process, and reduce uncertainty, is illustrated by comparing DO criteria between Arkansas and Louisiana. In Louisiana, a statewide DO criterion of 5 mg/L was initially established for all waters, including those naturally experiencing relatively low DO levels. Consequently, numerous water bodies were placed on the 303(d) list due to low DO. Many of these waters were delisted after the state applied the use attainability assessment (UAA) process to develop lower DO criteria for "dystrophic waters." Arkansas, in contrast, developed ecoregion-based DO criteria, ranging from 3 to 6 mg/L based on season and/or ecoregion, and had relatively few waters that were impaired due to DO exceedances.

WQS can incorporate physiography in various ways. For example, California has regional water quality boards that establish WQS. These regional boards, in effect, incorporate physiographic or ecoregional differences within the state. Texas establishes DO criteria in some parts of the state based on stream head slope as a way of incorporating physiography into its WQS.

Developing physiographic or ecoregion-specific criteria can potentially require significant state resources. However, even criteria that are based on relatively coarse classifications can significantly improve the accuracy of listing decisions. For example, Texas uses different approaches for assigning DO criteria in the western half of the state than in the eastern half.

5.2.1.2 Use Site-Specific WQS for Selected Water Bodies

Site-specific criteria determined through UAAs or other investigations (e.g., EPA 1994b) can significantly reduce the uncertainty associated with each site but at potentially significant cost. Ohio has established subcategories for site-specific water bodies, including those that are irretrievably impaired or manmade (e.g., drainage ditches). All the surveyed states and many other states have experience with UAAs. Some states have also performed UAAs on a regional basis (e.g., Edwards Aquifer in Texas) in addition to a site-specific basis.

5.2.1.3 Establish Seasonal or Wet Weather Criteria

The surveyed states did not, as a whole, distinguish between wet and dry conditions in WQS although other states do consider seasonal variation in setting WQS. Aquatic systems are dynamic; their physical, chemical, and biological constituents vary over time. To reduce some of the uncertainty in comparing fixed WQS with time-varying data, some states have also established variable WQS to protect designated uses during different periods. These periods might be related to seasons, storm events, or daily variability (Table 5-1). For example, Oregon allows water temperature exceedances during periods of extremely hot weather. Most states have seasonal limits for primary contract recreation with more stringent criteria on fecal coliform bacteria during the summer months than winter months. In addition, turbidity and suspended solids are expected to exceed dry weather or base-flow criteria during storm events, but these exceedances may not significantly impact some designated uses. The natural acidity of runoff from pine forests results in pH excursions below the 6.0 criterion in Arkansas and Mississippi streams without any impact on the biological community.

To address these situations, some states consider hydrologic variability or flow when evaluating data. For example, Kansas considers samples that are collected during high flows (upper 10th percentile) to be nonrepresentative, although they are required to use these sources for listing purposes. Water bodies that are listed as a result of nonrepresentative samples are given special consideration in the TMDL process. Similarly, for listing purposes, Oklahoma excludes turbidity data not collected during base-flow conditions.

5.2.1.4 Use Tiered WQS

Ohio uses a stratified or tiered approach in establishing designated uses to help reduce uncertainty in designated uses and associated criteria. For example, for aquatic life habitat, Ohio has multiple categories including warmwater, limited warmwater, exceptional warmwater, modified warmwater, seasonal salmonid, coldwater, and limited resource water. The distinctions among exceptional warmwater, limited warmwater, modified warmwater, and warmwater fisheries are related to their ability to support and maintain a balanced, integrated, adaptive community of warmwater aquatic organisms having a species composition, diversity, and functional organization comparable to the 25th percentile of the identified reference sites within specified ecoregions. For example, exceptional warmwater habitats have an aquatic community with attributes comparable to the 75th percentile of ecoregion reference sites. Limited warmwater habitats have individual site criteria that have been varied on a site-by-site basis. Modified warmwater are habitats that have had UAAs and been found incapable of supporting and maintaining biotic communities found in warmwater habitats because of irretrievable modifications of the physical habitat. Similar tiered categories are provided for other designated uses. Ohio has numeric biological criteria, but other numeric water quality criteria can also be modified to reflect the different designated uses.

5.2.1.5 Establish WQS Based on Magnitude, Frequency, and Duration

NRC (2001) recommended that all chemical criteria and some biological criteria be defined in terms of magnitude, frequency, and duration of exceedance. Magnitude represents the maximum value that can be tolerated over a specified period of time. This period of time represents the duration of exposure. The frequency component should be expressed as a number of permissible excursions in a specified period. While the concept has been proposed for some time, to our knowledge, no states have been able to implement criteria with all three components. Kansas has proposed incorporating frequency into its definition of critical periods, but has not incorporated magnitude, frequency, and duration into its WQS. Toxicity standards do have duration and frequency included in the definition, but magnitude is not explicitly considered. WERF anticipates funding a project in 2003 to develop approaches for incorporating magnitude, frequency, and duration into WQS.

5.2.1.6 Use Only Numeric Criteria or Establish Translators for Narrative Criteria

All states have numeric and narrative criteria for WQS. Many states, such as Mississippi and Arkansas, do not use narrative criteria for listing unless the potential pollutant or causative factor can be linked to numeric criteria. Water quality criteria fall into one of three general categories: biological (e.g., benthic invertebrate or fish community structure), chemical/physical (e.g., DO, temperature, pH, toxics), and toxicological (toxicity testing of ambient waters or effluents).

Biological Criteria. EPA (1990) defined biological criteria as numerical values or narrative expressions that describe the reference biological integrity of aquatic communities inhabiting water of a given designated aquatic life use. NRC (2001) noted that an attractive property of biological criteria is that they can provide a direct measure of aquatic life use attainment. Although biological assessments are useful for identifying biological impairments, they do not necessarily identify the causes of impairment, and further investigation procedures are required to identify stressors and lead to appropriate corrective measures (EPA 2000a). The development of numeric criteria for biological indicators and metrics can help reduce uncertainty related to narrative criteria for support of aquatic life designated uses.

Ohio has established numeric biological criteria for all aquatic life habitats in the state, and Mississippi is developing biological criteria (index of biotic integrity, or IBI) for its wadeable streams. Mississippi is considering retaining narrative criteria for the support of aquatic life, but has developed a calibrated IBI that may be used as a numeric translator. The IBI scores will be used as part of the assessment methodology rather than included in the WQS. As additional information becomes available to refine the biotic index, it can be modified without having to revise the WQS. Some states in the survey conducted for this study indicated a preference for retaining biological criteria as narrative rather than numeric criteria. These states preferred to use biological criteria as part of a weight-of-evidence approach in support of other types of information.

Chemical/Physical Criteria. The overwhelming majority of water quality criteria are physical or chemical. Criteria based on chemical or physical measurements have the advantage of being straightforward to sample and measure and are amenable to the kinds of quantitative analyses required for TMDL development. However, they provide a less direct measure of aquatic life use support (NRC 2001). In addition, chemical specific criteria have been developed for relatively few of the wide range of pollutants and mixtures that can be found in aquatic systems that can threaten aquatic life (Heber et al 1996).

Toxicological Criteria. Criteria based on toxicity can be useful complements to chemical specific criteria because they integrate interactions among complex mixtures of contaminants and measure a total toxic effect regardless of physical and chemical composition of a water body (Chapman 2000). However, like biological criteria, toxicity tests do not identify the cause of toxicity. Further investigation is necessary identify the cause(s) of toxicity and allow appropriate estimates of pollutant loadings and control measures. Toxicity testing can also be an effective tool for predicting instream impacts on aquatic life, provided appropriate consideration of exposure is made (Waller et al 1996; Dickson et al 1992; De Vlaming and Norberg-King, 1999; Diamond and Daley 2000). This issue is a contentious one. (See Hall and Giddings, 2000.)

5.2.1.7 Conduct Use Attainability Analyses (UAAs)

A UAA must be conducted when a state considers designation of a use less stringent than the fishable/swimmable goals of the Clean Water Act. The UAA process can lead to (1) confirming the appropriateness of the designated use and supporting criteria as water quality goals for the water body; (2) confirming the designated use and developing site-specific criteria to protect the use; or (3) adding or removing designated uses and developing appropriate criteria to protect the uses (WERF 1997).

These and similar procedures for modifying WQS (e.g., site-specific criteria development, water effects ratios, or recalculation procedures) can increase the likelihood of establishing WQS that are appropriate for the water body. Appropriate WQS in combination with adequate monitoring data and sound methodology should result in reduced uncertainty regarding the accuracy and appropriateness of attainment classifications. Other alternatives address the uncertainty associated with the water quality data or the comparison methodology used to compare the data with the standard.

5.2.2 Water Quality Data

All states have established procedures and protocols for reducing the uncertainty and increasing the representativeness of the water quality data collected through their monitoring programs. These approaches address:

- Monitoring program design;
- QA/QC requirements (through legislative bodies or policy);
- Evaluated information; and

- Third-party protocols (Table 5-2).

5.2.2.1 Monitoring Program Design

There are two parts to any monitoring program design: sampling design and response design (Olsen et al. 1999). Sampling design focuses on the number of sites and how sites are selected to ensure the site is representative of water body conditions. Response design focuses on what is measured, how frequently, where, and how it is analyzed to describe the water quality conditions in the water body. The design of both monitoring elements should be dictated by the specific objectives of the monitoring program and ultimate use of the information. In general, there has been greater attention given to the response design than the sampling design. Appropriate sampling design, however, can reduce the risk of inappropriate listings.

Monitoring Program Sampling Design. The two general approaches for sampling design are targeted site selection and probabilistic site selection (Olsen et al. 1999; Stevens 1994). Most monitoring networks have been designed using site selection based on the judgment of the investigators. For example, the site might have been selected to assess the effect of point source discharges into the receiving system, to characterize nonpoint source contributions from particular areas, to reflect the overall conditions of a watershed, or to meet other goals.

There is considerable guidance for targeted site selection (EPA 1986, 1996; Harmancioglu et al. 1999; Ward et al. 1997). An issue associated with targeted site selection is that these sites cannot be used to infer the condition of other sites. The information obtained in monitoring these sites is applicable only for that water body segment. For 303(d) purposes, this is satisfactory because the listing/delisting process focuses on individual water bodies. The condition of other sites is unknown. All states have targeted sites as part of their monitoring network.

Some states such as Maryland, Maine, South Carolina, and West Virginia have established probabilistic monitoring networks based on sample survey designs. Probabilistic sampling designs sample resources in direct proportion to their occurrence on the landscape. This permits information on the condition of individual sites to be used to estimate the condition of the population of water bodies (e.g., wadeable streams, lakes, estuarine area) within the state (Larson et al. 1994; Olsen et al. 1999; Stevens 1994; Macauley 1999). EPA (2002) recommends that state monitoring designs should allow the estimation of the percentage of impaired water bodies statewide to within ±10% at a 90% confidence level. Although probabilistic monitoring networks provide information useful for determining the proportion of the resource that is attaining or exceeding WQS (with known confidence intervals), they do not indicate the location of other sites estimated to have similar water quality conditions, but that were not sampled. Therefore, this information is useful for 305(b) assessments, but might not be as relevant for 303(d) purposes.

South Carolina uses a hybrid design of fixed stations along with probabilistic stations. Probabilistically selected sites provide information on the overall condition of state waters, including the proportion attaining WQS, with known confidence. A rotating basin approach to monitoring is used with fixed stations located at the mouth of each 11 digit watershed assessment unit. Data from these fixed sites are compared with WQS as part of the listing and delisting process.

Monitoring Program Response Design. The response design considers the specific variables to monitor, where at the site to monitor (including depth), how frequently to monitor, when to monitor, and how many samples are required to characterize the site accurately. Although response design is typically treated as a scientific or technical area, policy decisions play an integral role in monitoring program design, through the specification of data quality objectives (DQOs) supporting decision making. Obviously, the constituents incorporated in a monitoring program should relate to the water quality criteria (both numeric and narrative) and associated designated uses. All the states monitor constituents with numeric criteria and most have constituents that can be related to narrative criteria.

5.2.2.2 QA/QC Requirements

One of the primary approaches to reducing uncertainty in monitoring data is through quality assurance programs. All states have internally adopted QA/QC protocols for their monitored data, but they vary in how stringent their QA/QC requirements are for these and other data (GAO 2002). Data quality and quantity needs may vary according to the consequences of the resulting water quality decisions. Choosing the appropriate level of rigor is, in part, a policy decision. Some states, such as Florida, require that any data used in assessment activities must satisfy published QA/QC protocols.

Seven states have legislated data integrity laws that establish requirements for the quantity and/or quality of data that can be used in listing and delisting decisions (GAO 2002). These policies help reduce uncertainty in listing decisions by ensuring that data of unknown quality are not used. EPA has moved toward performance based measurement systems, instead of requiring specific techniques or procedures (EPA 1997). By specifying performance based criteria, the regulators and regulated community have the flexibility to use the techniques and procedures that are scientifically defensible, but also cost effective for the agency, organization, or industry.

5.2.2.3 Evaluated Information

Evaluated information is information based on modeling predictions or subjective assessment rather than quantitative data. All states use evaluative approaches or procedures for assessing waters that do not have monitoring data. However, most states do not list based solely on evaluative information. Rather, this information is used to target waters for monitoring. In the early 1990s, Mississippi used evaluative procedures (i.e., questionnaires and surveys) to identify waters that might be impaired from nonpoint sources as potential candidates for 319 projects. Numerous candidate water bodies were submitted in response to potential funding for 319 projects. However, subsequent litigation surrounding the listing/delisting and TMDL processes resulted in these water bodies being added to the 303(d) list even though no water quality data had been collected. More than 75% of the Mississippi listings resulted from this evaluative information. State 305(b) and 303(d) guidance should indicate that evaluative information can not be used as the sole source for listing, but might be used for targeting water bodies for monitoring.

5.2.2.4 Third-Party Protocols

Because uncertainty can be addressed by increasing sample size, volunteer or third-party data are used by some states to increase the quantity of data available for assessing the condition of their waters. The issues about using such data center on QA/QC. Massachusetts, for example, has guidance for volunteer monitoring to ensure that quality assurance procedures are followed during data collection and analysis so that these additional data can be used in the state assessment activities. Georgia and Utah include most data from universities, other state agencies, and federal agencies in their assessments. Few states use data for listing assessments if the quality of the data cannot be determined or is determined to be deficient. Some states, however, use this information to target additional monitoring if the data indicate the water body might be impaired.

5.2.3 Comparative Methodology

Some states have refined their approaches for comparing water quality data with WQS so that the uncertainty, if not reduced, is at least known and specified. One such approach is the establishment of statistically based methodology (e.g., incorporation of Type I or II error, use of formal statistical tests and procedures, establishment of minimum sample numbers). In other cases, states use verified evaluative procedures (Table 5-3) or use weight-of-evidence approaches. The subsections that follow describe these approaches in greater detail.

The approaches for comparing data with WQS for listing have varied from determining exceedances of WQS based on questionnaires, to the use of formal statistical procedures to assess the likelihood of exceedance (e.g., Mississippi, Florida). Delisting approaches have varied from delisting upon completion of a TMDL, to demonstrating through additional data collection and statistical analyses that WQS are not exceeded.

In comparing criteria with monitoring data, the criteria are assumed to be absolute, with no variance around the standard. However, because of the uncertainties in WQS and water quality data noted above, it is reasonable to treat the assessment process as a statistical decision problem (Smith et al. 2001). EPA guidance (EPA 1997) recognizes these uncertainties and the statistical aspects of the listing/delisting process. The EPA guidelines recommend that if more than 10% of the samples collected from the water body exceed the criteria, the water body should be listed. Several states (Mississippi, Arkansas, Louisiana) use a percentage approach, with 10% as the threshold between fully supporting and partially supporting designated uses. Water bodies are partially supporting designated uses if between 10% and 25% of the samples exceed criteria. If greater than 25% of the samples exceed the criteria, the water body is classified as not supporting designated uses and is placed on the 303(d) list. This quasi-statistical approach is termed the "raw score" test (Smith et al. 2001).

5.2.3.1 Establish Statistically Based Methodology

Incorporation of Type I or Type II Error. In general, statistical approaches are used to test the statement or hypothesis that the water body is not impaired (i.e., the null hypothesis). The two types of errors associated with statistical tests — Type I and Type II errors — were described in Section 4.3.2. A disadvantage of using a raw score test (described in the previous section) is that, because Type I and Type II errors are not explicitly addressed, their magnitude is unknown. This means that the probability of incorrect classifications is unknown. In addition, the raw score test has been shown to have high Type I error rates (Smith et al. 2001). This means that using the raw score test will tend to overestimate the number of water bodies that are impaired. If the state policy or philosophy is conservative, that is, if the policy is to list any water body suspected of being impaired, then the tolerance for Type I errors may be relatively high. However, unless Type I and Type II error rates are explicitly addressed, there is no way for the state to choose a methodology that is consistent with its philosophy or policy.

Use of Formal Statistical Tests and Procedures. Two statistical tests that have been compared with the raw score test are the binomial test and the Bayesian binomial test (Smith et al. 2001). Both of these tests control for Type I error rates, whereas the raw score test does not. Florida has adopted the binomial test and incorporated it into its Water Quality Regulations for both listing and delisting. In its regulations, Florida specifies the Type I error rate, the confidence interval, and the number of violations of WQS that are permitted as a function of sample size before the water body is classified as exceeding the standards. (See Tables 5-5 and 5-6.)

The Bayesian binomial test is another test that controls for Type I error rates, but it incorporates prior opinions (information) about the water body into the test. Having a strong prior opinion that there is no impairment in the water body produces results that have lower Type I error rates than the binomial test does (Smith et al. 2001). Both the binomial and the Bayesian binomial tests have higher Type II error rates than the raw score test, again reflecting the complementary nature of Type I and Type II errors. However, as the sample size increases, Type II errors converge toward zero (Smith et al. 2001).

While the discussion above appears to be strictly technical, the specification of the desired Type I error rate, confidence level, and number of exceedances that constitute impairment are policy, not technical, decisions. This is because the magnitude of the selected Type I error rate and the number of exceedances that constitute impairment reflect the state's tolerance for incorrectly classifying water bodies as impaired. Once these two issues are resolved, determining the minimum sample size to achieve these objectives is straightforward. It is also technically possible to specify Type II error rates. This would represent a different philosophical or policy approach that focuses on the rate at which impaired water bodies are incorrectly classified as unimpaired. Once the error rate, confidence level, and minimum sample size are specified, the number of violations that can be tolerated before WQS are said to be exceeded can be estimated as a function of sample size. Florida, for example, has different Type I error rates, confidence levels and minimum sample size for three lists: a planning list, a verified [or 303(d)] list, and delisting (Tables 5.5 through 5.7).

Establish Minimum Sample Numbers. As indicated above, with the specification of the Type I error rate, Type II error rates vary according to the sample size and measurement uncertainty. Several states, including Florida, Mississippi, Texas, and Washington, have specified a minimum number of samples for assessing water bodies. The minimum number of samples is typically around 10, although this number is not necessarily based on statistical considerations. Florida uses a minimum of 10 samples only for its planning list, with a minimum of 20 samples required for assessing water bodies for its verified list (see discussion of Florida lists above). Examples of the Type I error, confidence interval and minimum sample size are shown for Florida's planning list (Table 5-5) and verified list (Table 5-6).

When the sample size is greater than 20, the binomial, Bayesian binomial, or similar statistical test should replace the raw score test because these tests better control Type I and Type II error rates. When the sample size is less than 20, the Bayesian binomial test should be considered. The binomial test does not control error rates as well as the Bayesian binomial test with sufficient prior information (Smith et al. 2001).

Florida Example. An example of a statistically based comparison methodology is found in Florida's water quality standards (FDEP 2002). The state has established that a water body is impaired when an applicable water quality criterion is exceeded in 10% or more of samples. The state developed a statistically based procedure to determine when this exceedance rate is reached.

The state has explicitly addressed Type I error rate protection by establishing a "Planning List" and a "Verified List." The Planning List identifies waters that are potentially impaired but require additional information. To be placed on this list, the number of exceedances must provide 80% confidence that the actual exceedance rate is greater than or equal to 10%. A minimum of 10 samples is required for this determination. The relationship between the minimum number of samples and the minimum number of exceedances is shown in Table 5-5. For example, if 20 samples are available and three exceedances are observed, the water body will not be placed on the Planning List, even though more than 10% of the samples exceed the criterion (Table 5-5). However, with 20 samples and four exceedances, the water body would be placed on the Planning List. Essentially, this approach protects Type I error at $P = .2$, allowing the state to identify potentially impaired waters with a relatively high tolerance for Type I errors. Water bodies that are placed on the Planning List are targeted for additional monitoring.

The Verified List identifies water bodies showing stronger evidence of impairment. These water bodies require TMDL development. To be placed on the Verified List, the number of exceedances must provide 90% confidence that the actual exceedance rate is greater than or equal to 10%. A minimum of 20 samples is required for this determination. The relationship between the minimum number of samples and the minimum number of exceedances is shown in Table 5-6. Because of the higher level of confidence adopted for the Verified List, more exceedances are required to place a water body on the Verified List than on the Planning List. For example, with 20 samples and four exceedances, a water body would be placed on the Planning List but not the Verified List. With five exceedances, however, the water body would be placed on the Verified List and scheduled for TMDL development (Table 5-6). The Verified List protects Type I error at a $P = .1$ allowing the state to identify impaired waters with a lower tolerance for Type I errors.

Table 5-5. State of Florida Planning List. Table entries indicate minimum number of measured exceedances (and corresponding required sample sizes) to place a water body on the Planning List with at least 80% confidence that the actual exceedance is greater than or equal to 10%.

Sample Sizes		Are listed if they have at least this # of exceedances
From	To	
10	15	3
16	23	4
24	31	5
32	39	6
40	47	7
48	56	8
57	65	9
66	73	10
74	82	11
83	91	12
92	100	13
101	109	14
110	118	15
119	126	16
127	136	17
137	145	18
146	154	19
155	163	20
164	172	21
173	181	22
182	190	23
191	199	24
200	208	25
209	218	26
219	227	27
228	236	28
237	245	29

Sample Sizes		Are listed if they have at least this # of exceedances
From	To	
246	255	30
256	264	31
265	273	32
274	282	33
283	292	34
293	301	35
302	310	36
311	320	37
321	329	38
330	338	39
339	348	40
349	357	41
358	367	42
368	376	43
377	385	44
386	395	45
396	404	46
405	414	47
415	423	48
424	432	49
433	442	50
443	451	51
452	461	52
462	470	53
471	480	54
481	489	55
490	499	56
500	500	57

Table 5-6. State of Florida Verified List. Table entries indicate minimum number of measured exceedances (and corresponding required sample sizes) to place a water body on the Verified List with at least 90% confidence that the actual exceedance is greater than or equal to 10%.

Sample Sizes		Are listed if they have at least this # of exceedances
From	To	
10	11	3
12	18	4
20	25	5
26	32	6
33	40	7
41	47	8
48	55	9
56	63	10
64	71	11
72	79	12
80	88	13
89	96	14
97	104	15
105	113	16
114	121	17
122	130	18
131	138	19
139	147	20
148	156	21
157	164	22
165	173	23
174	182	24
183	191	25
192	199	26
200	208	27
209	217	28
218	226	29
227	235	30
236	244	31

Sample Sizes		Are listed if they have at least this # of exceedances
From	To	
245	253	32
254	262	33
263	270	34
271	279	35
280	288	36
289	297	37
298	306	38
307	315	39
316	324	40
325	333	41
334	343	42
344	352	43
353	361	44
362	370	45
371	379	46
380	388	47
389	397	48
398	405	49
407	415	50
416	424	51
425	434	52
435	443	53
444	452	54
453	461	55
42	470	56
471	479	57
480	489	58
490	498	59
499	500	60

The procedure also provides a means for delisting water bodies. Interestingly, the procedure requires stronger evidence and more samples than the corresponding listing procedure. The minimum sample size is 30. With this sample size no exceedances are allowed for delisting (Table 5-7). This difference between listing and delisting requirements is explained in the support documentation for Florida's statistical listing and delisting procedure (Lin et al. 2000) as follows.

> "Requiring 'more samples' for delisting than for listing an impaired water at a comparable level of confidence seems somewhat puzzling to many readers, but it is strictly a matter of statistical theory. For example, suppose the agency decides that if *P* [the observed proportion of exceedances] is shown to be greater than .1 then the water will be listed as impaired. Assuming a null hypothesis of $P = .1$, the variance of each observation is .1 _ 0.9 = 0.09. Now suppose the water body is listed as a result of a random sample. Then the agency will assume a null hypothesis of 0.2 for the purpose of testing for delisting. Now the variance of each observation is 0.2 _ 0.8 = 0.16. Since the sample size necessary *to create the same level of confidence* for the estimation of *P* is roughly inversely proportional to the variance of an observation in a random sample, it will take more observations to provide the same standard of proof when $P = .2$ as when $P = .1$. Consequently, it is not possible to use the same sample size to list and delist an impaired water body reach at the same level of confidence using the 10% exceedance method for both listing and delisting."

Comments by professional organizations (Table 3-2) and the results of the state survey (Section 4.3.5, Question 3) indicated a preference that the requirements for delisting be similar to those for listing. This situation could pose a dilemma for states that wish to adopt a statistically based comparison methodology similar to Florida's but also prefer that listing and delisting have similar standards of proof. This situation could possibly be addressed through alternative definitions of impairment versus nonimpairment.

Table 5-7. State of Florida Delisting Criteria. Table entries indicate the maximum number of measured exceedances allowable (and corresponding sample sizes) to delist with at least 90% confidence that the actual exceedance rate is less than 10%.

Sample Sizes		Maximum # of exceedances allowable for delisting
From	To	
30	37	0
38	51	1
52	64	2
65	77	3
78	90	4
91	103	5
104	115	6
115	127	7
128	139	8
140	151	9
152	163	10
164	174	11
175	186	12
187	198	13
199	209	14
210	221	15
222	232	16
233	244	17
245	255	18
256	266	19
267	278	20
279	289	21

Sample Sizes		Maximum # of exceedances allowable for delisting
From	To	
290	300	22
301	311	23
312	323	24
324	334	25
335	345	26
345	356	27
357	367	28
368	378	29
379	389	30
390	401	31
402	412	32
413	423	33
424	434	34
435	445	35
446	456	36
457	467	37
468	478	38
479	489	39
490	500	40

5.2.3.2 Use Verified Evaluative Procedures

Evaluative procedures such as empirical, steady-state, and dynamic models have been used to assess waters that have not been monitored. Evaluative procedures, however, are not without uncertainty. Using models without verification that their predictions are within the desired accuracy and precision for use in assessment can result in misclassification of water bodies without any accompanying estimate of uncertainty. These evaluative procedures might provide information for targeting water bodies for monitoring, but are not recommended for listing decisions. Few, if any states, have explicitly collected data to verify their evaluative procedures.

5.2.3.3 Weight of Evidence

Weight of evidence is a term that is widely used in environmental assessment. During the state surveys, officials indicated a desire for a flexible, weight of evidence approach to listing. Although some specific approaches have been proposed (e.g., the "sediment triad"; Chapman 1986) there is no generally accepted procedure or framework for applying a weight-of-evidence approach to environmental assessment.

Weight of evidence can refer to agreement between different lines of evidence or a preponderance of evidence. Examples of circumstances in which a weight-of-evidence approach can be used are as follows:

- *To resolve instances where different data sets (e.g., biological versus chemical) indicate different conclusions regarding impairment:* For example, Montana's assessment methodology uses a weight-of-evidence test to indicate agreement or corroboration among data. The state uses this test to make aquatic life use support decisions when there is sufficient and credible data available for all three data categories recognized by the state (biological, physical/habitat, chemical/toxicity) and at least two biological assemblages were assessed. A determination of "fully supporting" requires all data categories to indicate that the water body is unimpaired or no more than one category indicating moderate impairment. By contrast, a determination of "not supporting" requires one or more data categories indicating moderate impairment in combination with a separate category indicating severe impairment.
- *To address water quality issues that are not easily assessed using a single parameter or criterion:* North Carolina applies a weight-of-evidence approach to assess eutrophication in lakes. The approach uses the following sources of information, which are listed here in no particular order: multiple quantitative water quality variables (DO, chlorophyll *a*), third-party reports, analysis of water quality complaints, algal bloom reports, macrophyte observations, reports from water treatment plant operators, reports from lake associations, fish kill reports, taste and odor observations, aesthetic complaints, frequency of noxious algal activity, and reports or observations from state agencies. An example of this approach is provided in North Carolina's listing methodology. In that example, a lake exhibited slight exceedances of the state's chlorophyll *a* standard. However, because there were no supporting indications of lake eutrophication (e.g., nuisance algal blooms, taste and odor problems, citizen complaints), the lake was determined to be meeting its designated uses and was not considered to be impaired.
- *To support narrative criteria:* Arkansas requires that determinations of impairment based on narrative criteria be supported by exceedances of numeric criteria or indications of impaired aquatic life use as determined through site-specific biological assessments.
- *To support (or not support) indications of slight or moderate impairment:* See preceding examples.

Some states (Arizona, North Carolina) use weighting systems to resolve instances where different data types (e.g., chemical vs. biological) provide conflicting evidence of impairment. These systems are informal and do not involve explicit weighting or scoring factors. For example, North Carolina weights different types of data according to the following hierarchy, from highest to lowest weight:

1. benthic classification/fish community structure;
2. chemical/physical data;
3. monitored data > 5 years old; and
4. compliance/toxicity data.

The North Carolina methodology stresses that this hierarchy is only a general guideline and that a particular assessment may vary from this based on best professional judgment and site-specific conditions.

The fundamental difference between weight of evidence and *weighting* is that weight of evidence involves evaluating whether or not different lines of evidence are consistent or support a particular conclusion. Greater consistency among data types might be expected with severe impairment than with marginal impairment although it is with marginal impairment that weight of evidence may be most needed. Applying a weight-of-evidence approach does not *require* multiple lines of evidence. Strong evidence of impairment from a single line of evidence can be sufficient to classify a water body as impaired. Weighting, by contrast, involves assigning different levels of importance to different data sets based on the type of data or other considerations (e.g., age or amount of data).

No generally accepted guidance exists for weighting or for applying weight of evidence. Rules and guidelines for assigning weight or applying weight of evidence are essentially policy decisions.

5.2.4 List Format and Content

In addition to the alternatives to reduce uncertainty discussed above, the list format and content also can provide a means for managing uncertainty. Some of the approaches that can be used to reduce the uncertainty in categorizing or classifying water bodies include:

- Use the EPA 2002 Integrated Water Quality Monitoring and Assessment Report Guidance (EPA Listing Guidance);
- Develop planning and verified lists; and
- Delist using EPA Integrated Assessment Guidance, specific delisting criteria, phased TMDLs, or adaptive management (Table 5-4).

Each of these approaches for reducing uncertainty associated with categorizing water bodies are discussed in greater detail in the subsections that follow.

5.2.4.1 Use EPA 2002 Listing Guidance

Recently, EPA provided guidance on an alternative approach for classifying water bodies into five categories, only one of which is the 303(d) list and thus requires a TMDL (EPA 2001b). The EPA categories are shown in Table 5-8. A diagram summarizing how water bodies are placed into these five categories is provided in Figure 5-2.

Table 5-8. Classification Scheme for Water Bodies Based on EPA 2002 Listing Guidance

EPA Category	Description
Category 1	Waters that are attaining the water quality standard and where no use is threatened. States should consider additional monitoring of these waters to document their compliance status.
Category 2	Waters that are attaining some designated uses and where no use is threatened, but for which there are insufficient or no data or information to determine whether other designated uses are attained or threatened. Additional monitoring should be conducted to obtain the data necessary for use attainment decisions.
Category 3	Waters for which there is insufficient information or no data or information to determine whether any designated use is attained or threatened. Additional monitoring should be conducted to obtain necessary data.
Category 4	Waters that are impaired or threatened for one or more designated uses, but for which the development of a TMDL is not required for one of the following reasons: a. A TMDL has already been developed that when implemented is expected to result in full attainment of WQS. If more than one pollutant is causing impairment the water body will remain in category 5 until all TMDLs are completed and approved by EPA. Additional monitoring should be conducted to assess the effectiveness of TMDL implementation. b. Other pollution control requirements are reasonably expected to result in attainment of standards in the near future. (EPA expects that the requirements referred to here are specifically applicable to the water quality problem at issue. Follow-up monitoring must be scheduled to verify that the standard is obtained as expected). c. The impairment is not caused by a pollutant. Additional monitoring should be conducted to confirm that there is no pollutant caused impairment and to support management actions necessary to address the cause(s) of impairment.
Category 5	Waters that are impaired by one or more pollutants, and for which development of a TMDL is required. This category constitutes the Section 303(d) list of water bodies impaired or threatened by pollutants for which one or more TMDLs are required.

This guidance implicitly acknowledges the uncertainty that surrounds the listing/delisting process and provides alternative categories for water bodies to accommodate this uncertainty. One of the issues that has arisen related to this listing guidance relates to the treatment of biological data. In a 26 March 2002 memorandum, EPA stated that water bodies impaired based on biological data and information, for which the cause of the impairment is unknown, may not be placed in Category 3. If the water body is impaired by pollution, it should be placed in Category 4C. When existing and readily available data and information are sufficient to determine that a pollutant has caused, is suspected of causing, or is projected to cause the impairment, the water body should be listed in Category 5. However, there is no category if the cause of impairment is unknown, and the biological data indicate the water body is impaired. Without a pollutant cause, a TMDL cannot be conducted. Mississippi has proposed a fourth subcategory under Category 4 — 4D. Category 4D would apply when biological data or information indicate the water body is impaired, but the cause of the impairment is unknown. This water body would require follow up monitoring to determine the cause of the impairment.

5.2.4.2 Develop Planning and Verified Lists

As explained in the earlier example, Florida has two lists for its water bodies. The Planning List represents water bodies that are suspected to be impaired but have not been fully verified. This list has a minimum sample size of 10, a Type I error rate of 0.1, and a confidence level of 80%. Subsequent verification of these water bodies includes the collection of additional data, if needed, with a minimum sample size of 20, a Type I error rate of 0.1, and a confidence level of 90%. Water bodies that are verified to exceed WQS are placed on the verified or 303(d) list for TMDLs. The NRC report (2001) also recommended two lists, a preliminary list and a 303(d) list, similar to that used by Florida.

5.2.4.3 Delisting

Delisting has not received the attention that listing has, in part, because of the litigation requiring TMDLs for listed waters. Regulatory restrictions on the uses of listed waters, however, dictate that consideration also be given to delisting. The uncertainty on attaining WQS once implementation of required point and nonpoint source reductions is initiated is as great, if not greater, than the uncertainties associated with listing waters.

EPA (2000b) provides very general guidance for delisting. The 28 April 2000 memorandum states that, "The existing regulations require states, at the request of the Regional Administrator, to demonstrate good cause for not including water bodies on their lists. Good cause includes, but is not limited to, more recent and accurate data, more sophisticated water quality modeling, flaws in the original analysis that led to the water body being listed or changes in conditions, e.g., new control equipment or elimination of discharges". California has provided additional criteria for delisting in its published listing methodology (Section 4.3.4).

Florida also provided guidance on delisting water bodies. To delist a water body requires a minimum of 30 samples, a Type I error rate of 0.1 and a confidence level of 90% (Table 5-7). The delisting requirements are more stringent than the listing requirements. Comments to EPA on the proposed rule making and the state survey results from this study all recommended that the delisting criteria be no more stringent than the listing criteria.

EPA 2002 Listing Guidance indicates that a water body is reclassified as Category 4A once the TMDL has been completed. Because Category 5 is the only category associated with the 303(d) list, the *de facto* conclusion is that a water body is delisted when it is no longer in Category 5.

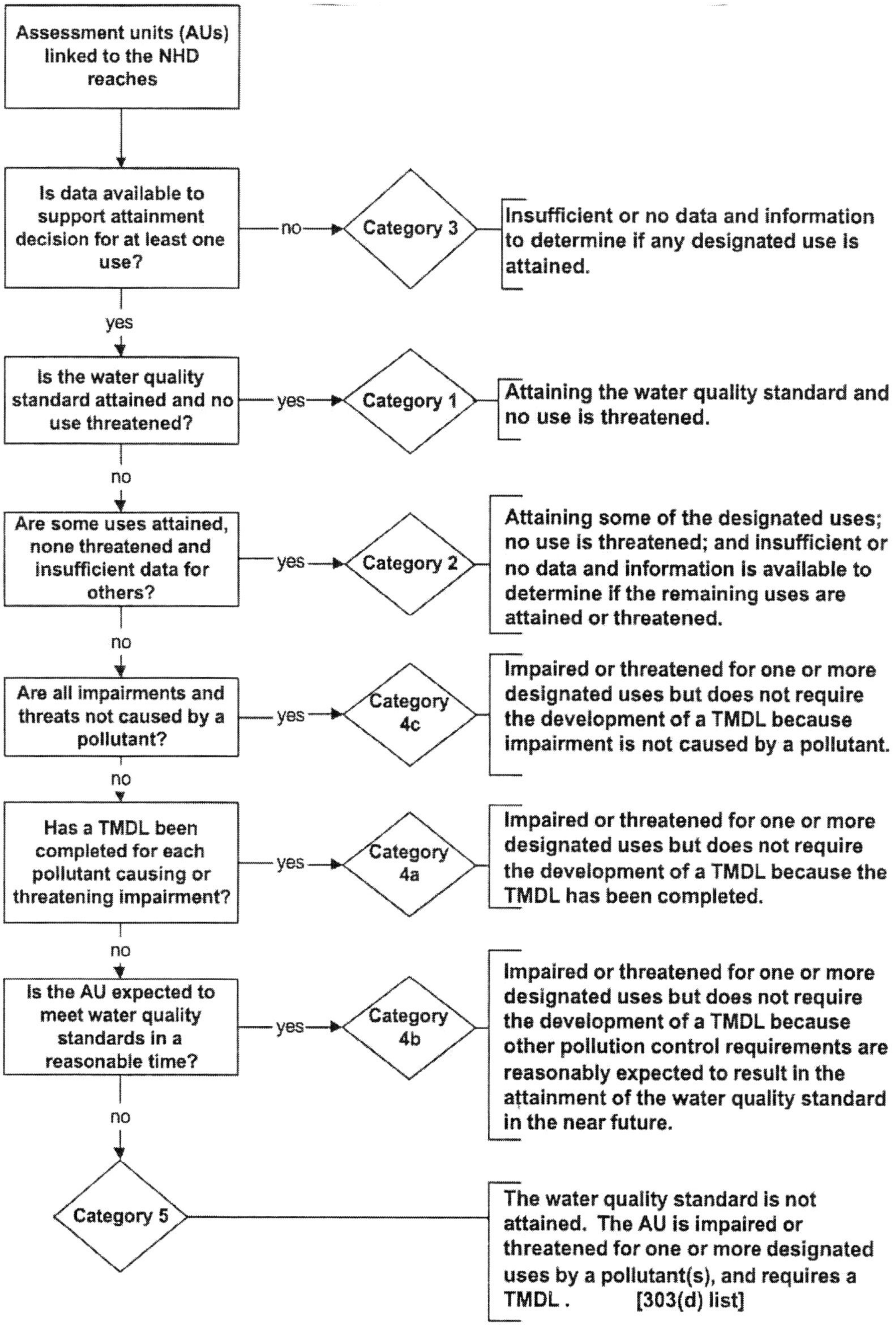

Figure 5-2. Diagram Summarizing EPA Guidance for the Water Bodies in the Five Listing Categories Described in EPA Listing Guidance (EPA 2001b).

The following sections offer examples of changes in state approaches that have or are expected to result in delisted water bodies.

Changes in Water Quality Criteria: DO in Louisiana Streams. The statewide DO criterion of 5 or 4 mg/L was amended to address concerns that DO criteria did not accurately reflect natural conditions in many of Louisiana's water bodies. The 2000 Louisiana 305(b) reports states that "To get results more representative of Louisiana waters, the year 2000 assessment process was modified slightly from assessment procedures utilized in earlier reports. Previously, dissolved oxygen data from water bodies that had not been individually studied was compared to a single criterion (usually 5.0 ppm for inland waters or 4.0 for estuaries). The water body was considered not fully supporting of the designated use of fish and wildlife propagation (FWP) if more than 10% of the data points fell below the criterion. In Louisiana, however, many water bodies fluctuate naturally to lower levels, and dissolved oxygen data is inherently broad. It has, therefore, become apparent that a one-value comparison does not allow for a good representation of Louisiana water bodies. To allow for natural fluctuations in DO data, a two-value assessment procedure was utilized in the year 2000 assessment....A minimum value and the median value were utilized (i.e., not supporting if >10% do not meet minimum of 3.0 ppm or median value is < 5.0 ppm). In this manner, water bodies were checked for sufficient DO to sustain aquatic life, yet were allowed to exhibit natural characteristics of Louisiana waters" (LDEQ 2000). This change resulted in the Louisiana and Region 6 renaming (delisting) for DO impairment, especially for water bodies considered to be "dystrophic."

Changes in Water Quality Criteria and Additional Monitoring Data: Biological Impairment and Pathogens in Mississippi Streams. Many water bodies placed on the Mississippi 1998 303(d) list were based on evaluations (i.e., a survey questionnaire) that included little or no assessment data. The state has since developed an index of biotic integrity (IBI), which will be the primary approach for assessing aquatic life impairment. Preliminary findings indicate a large number of water bodies will be removed from the 303(d) list during the next listing cycle.

In addition, many Mississippi water bodies were classified as impaired due to pathogens on the 1998 303(d) list. Many of these listings were based on one or few data points. Mississippi has since modified its pathogen assessment data requirements to require five measurements collected during two 30-day periods during the contact and noncontact seasons for at least 2 years (a total of 20 data points). This change is expected to result in some water bodies being reclassified from impaired to unimpaired but also will likely result in reclassification of other water bodies from unimpaired to impaired status.

Collection of More Accurate Data: Metals in Louisiana Streams. To address concerns that sample contamination may have resulted in inaccurate estimates of ambient metal concentrations, the state conducted a pilot study to examine the effects of "clean sampling" protocols on estimates of ambient metal concentrations. The 2000 Louisiana 305(b) reports states that "Due to closer scrutiny of metals criteria for water quality and the increased sampling of fish for the mercury contamination study, the suspected cause of metals, as a group, accounted for the most impaired stream miles. The suspected cause category of metals encompasses arsenic, cadmium, copper, lead, and mercury. A total of 4471 miles of streams experienced metals criteria exceedances or fish consumption advisories due to mercury. However, much of the impairment due to metals criteria exceedances may be the result of sample contamination. An investigation conducted in 1999 used clean sampling and analysis techniques to determine the exact nature of these exceedances. Results of this study demonstrated that all but one of the test water bodies did not have metals criteria exceedances. As a result, water bodies currently showing metals criteria exceedances will have this assessment confirmed prior to any form of TMDL development."

5.3 Comprehensive Approach

Many options used by different states can be combined or nested to provide a comprehensive approach for navigating the listing/delisting process and reducing uncertainty associated with determining attainment or nonattainment of WQS. This section discusses this comprehensive approach.

5.3.1 Water Quality Standards

In a comprehensive approach to WQS, a tiered approach to developing WQS may be supplemented with seasonal or temporal criteria and translators for narrative criteria. Under this approach, WQS would include considerations of:

- Water body type;
- Physiography or ecoregion;
- Site-specific criteria established through UAAs or similar procedures;
- Seasonal and temporal criteria; and
- Translators to numeric values for narrative criteria.

Ohio's approach to WQS includes many of these elements of a comprehensive approach. Whereas the triennial WQS review occurs every 3 years, a critical review of WQS as part of a rotating basin or similar monitoring schedule would provide continuous feedback on the applicability of designated uses and water quality criteria for water bodies throughout the state. This approach could also identify water bodies where modifications of uses and/or criteria, rather than TMDLs, might be appropriate.

5.3.2 Data Acquisition

Nested monitoring designs, considering both probabilistic and targeted sites, provide greater insight into water quality within the state, satisfying both the 305(b) and 303(d) requirements. This nested approach considers both the sampling and the response design. South Carolina has implemented a nested program for both freshwater and estuarine systems. QA requirements for state water quality monitoring programs have been established in every state. QA requirements (including sample size requirements) also need to be established for all data used in integrated water quality monitoring and assessment activities. Monitoring programs should be designed and implemented to permit an assessment of the waters of the state. If evaluative procedures are used as part of the assessment, data needed to verify these procedures should be collected as part of the monitoring effort.

5.3.3 Comparative Methodology

Both Type I error rates and minimum sample numbers should be explicitly stated, which implicitly specifies Type II error rates. Florida, for example, has specified a minimum sample size of 10 for the planning list, 20 for the verified list, and 30 to delist water bodies. The timeframe over which these samples can be collected should also be identified (e.g., 20 samples collected over 2 years, 3 years, and so forth) The formal statistical test to be used in comparing the WQS with the data should be stated. Specifying the test, which might be the binomial, Bayesian binomial, interval, or similar test, provides additional information on any required distributional attributes of the data, error rate characteristics of the test, and similar information on the likelihood or probability of misclassification of water bodies.

No unverified evaluative procedures should be used for assessing water bodies. Verified procedures should not be used outside the range of water body attributes used for its development and verification. Evaluative procedures should not be used for listing or delisting decisions, but they may be used to target water bodies for additional monitoring. The methodology for applying evaluative procedures can be part of the states' continuous planning process (CPP).

5.3.4 List Content and Format

The 2002 EPA Listing Guidance has been revised for developing a 303(d) List. Specification of Type I errors and minimum sample size permits an estimate of the probability of misclassification for Categories 1 and 2. Category 3 can be quantitatively defined through specification of the minimum sample size. For example, if the minimum sample size specified by the state's comparative methodology (Section 5.3.3) were not available to assess a water body, then it would be classified as having insufficient information and placed in Category 3. With the exclusion of Category 3 for biological impairment with unknown cause; however, adding another subcategory to Category 4 (e.g., Category 4D: "Biological impairment, but cause of impairment unknown. Additional monitoring required.") may be a way to resolve this issue. Toxicity of unknown cause could also be included in Category 4D. Although Category 5 represents the 303(d) list, there is no explicit guidance for delisting. If a water body was previously on the 303(d) list and a TMDL was completed, it then moves to Category 4A. Following the attainment of WQS, the water body moves to Category 1.

Additional guidance is required for phased TMDLs and adaptive management. If phased TMDLs were also included in Category 4A, with the understanding that additional monitoring would be required, then phased TMDLs would have similar status to completed TMDLs. Long timeframes are associated with implementation and effectiveness of best management practices (BMPs) for nonpoint sources. For example, a 20-year timeframe was associated with attaining nonpoint source reductions for the Chesapeake Bay, and decades are associated with reducing groundwater nutrient concentrations in the Snake River–Hells Canyon nutrient TMDL. Phased TMDLs and adaptive management permit progress toward WQS attainment and reduce the restrictions on the use of 303(d) listed water bodies.

CHAPTER 6.0

Conclusions and Recommendations

The ultimate goal of the Clean Water Act is to ensure that the designated uses of the waters of the United States are protected and sustained. The TMDL program is an integral part of this desired goal. The TMDL program, in conjunction with watershed planning and management, provides an approach for ensuring that the WQS are being attained in the nation's waters. The project team has documented several options that can be incorporated in the listing and delisting process to permit the desired goal of clean water to be attained, while more precisely targeting the TMDL process to only those waters that are not attaining WQS. This report presents a number of alternatives and options that are being used by various states to reduce uncertainty in the accuracy of listing decisions and to navigate through the listing-delisting process. A minimum set of activities was suggested for all states. States can undertake a comprehensive approach by incorporating those alternatives that are consistent with state policies and time and resource constraints.

6.1 State Survey Results

Analysis of the state survey results led to the following conclusions:

- Surveyed states recognize that physiography influences designated uses and criteria, but there were significant differences among states in how this variation was considered in WQS.
- Surveyed states generally do not consider hydrologic variability in establishing water quality criteria.
- With the exception of a relatively limited number of UAAs, surveyed states have not explicitly determined that designated uses are attainable.
- All surveyed states considered existing criteria protective of designated uses, but the rationale for developing criteria varied from site specific determinations to adopting EPA national criteria.
- Narrative criteria are useful for problem identification, but listing decisions should be made based on monitoring or site specific data.
- No surveyed states used a formal statistical approach to compare data with criteria.
- Evaluative procedures generally have been tested only through special studies or informally in the states surveyed.

- All states surveyed have formal QA/QC programs for their monitoring data and all require QA reviews for extant data.
- None of the states surveyed use formal statistical approaches for delisting.

The states surveyed recognized the need for site-specific criteria and suggested using (1) site-specific approaches, (2) existing conditions, (3) expert panels, as well as (4) development of criteria as part of the TMDL process. The state representatives, in general, believed that the listing process would be more reliable and useful if it were more flexible and based on additional data of higher quality. Moreover, the representatives expressed a need for more technical guidance from EPA not only on the listing process but also on delisting based on data.

6.2 Recommendations

6.2.1 Reducing Uncertainty in the Listing/Delisting Process

Based on the results of the state surveys, selected review of approaches in other states, comments provided on the 1999 proposed EPA rule, the NRC (2001) report, published literature, and the investigators' experience, it was clear that technical and scientific uncertainty is imbedded in all components of the listing/delisting process (i.e., WQS, water quality data acquisition, comparative methodology, content and format of the 303(d) list). These sources of uncertainty can result in inappropriate or inaccurate assessments regarding the attainment or nonattainment of WQS. Furthermore, the project team concluded that a number of options and alternatives to reduce uncertainty in the listing and delisting process are being used by states or have been published in the literature. A comprehensive approach should be implemented by adopting as many such alternatives as state policies, time, and resources permit.

Among the options and alternatives for reducing uncertainty and for navigating the listing/delisting process are these, which are organized along the lines of the four major elements of the process itself:

1. WQS
   - Establish WQS based on physiography, ecoregions, or water body type;
   - Use site-specific WQS for selected water bodies;
   - Establish seasonal or wet weather WQS;
   - Use tiered WQS;
   - Establish WQS based on magnitude, frequency, and duration;
   - Use only numeric criteria or establish translators for narrative criteria; and
   - Conduct use attainability analyses (UAA).

2. Water Quality Data Acquisition
   - Implement statistically designed monitoring program;
   - Develop appropriate response design;
   - Set QA/QC requirements (through legislative bodies or policy);
   - Use evaluated information; and
   - Establish third-party protocols.

3. Comparative Methodology
   - Establish statistically based methodology;
   - Use verified evaluative procedures; and
   - Utilize weight-of-evidence approach.

4. List Content and Format
   - Use the EPA 2002 Integrated Water Quality Monitoring and Assessment Report Guidance (EPA Listing Guidance);
   - Develop planning and verified lists; and
   - Delist using EPA Integrated Assessment Guidance, specific delisting criteria, phased TMDLs, and adaptive management approaches.

6.2.2 Improving the Process: Minimum Activities for the States

A minimum set of activities should be considered by all states as part of the listing/delisting process:

- Critically review the appropriateness of WQS, including revisions since the last assessment as well as revisions proposed for the next triennial review or legislative session.
- Establish QA/QC requirements for all data used as part of the listing and delisting process.
- Explicitly indicate a preference for controlling either Type I or Type II errors, or state the Type I error rate.
- Specify the confidence level (error rate) desired (permitted) in the tests comparing the data with the WQS.
- Adopt the List Content and Format categories in the EPA Listing Guidance.

The TMDL program is an important part of the effort to protect and sustain the waters of the United States. The listing and delisting alternatives offered here should assist the states as they navigate the TMDL process.

APPENDIX A

SUMMARY OF STATES' SURVEY RESPONSES

Number	Question	SUMMARY OF ALL EIGHT STATES
2.0 Water Quality Standards/Criteria and Designated Uses		
2.1.0	Are WQS based on ecoregion or physiographic criteria that incorporate factors related to natural assimilative capacity of the waterbodies?	No (IN,RI);Yes (MI,MA, AR); Ecoregion (RI); Seasonal criteria may apply where UAAs were conducted (LA); N/A (OR); WQS set by regional boards (CA)
2.1.1	Are WQS uniform for the entire state	Yes (RI, OR) ; Great Lakes Basin vs. rest of state (IN) ; Vary by ecoregion (AR); Vary by waterbody type (MI, LA); Vary by classes and designated use (MA) ; Vary per regional boards (CA)
2.1.2	Which parameters are uniform	All are uniform except Temp and DO (IN,MI[1]); All (RI,OR); narrative for oil & grease, taste & odor, color, turbidity, and solids are the same for all classes (MA); Fecal coliform (AR); DO (within categories) and bacteria (LA); WQS for hot springs and any parameters covered by California Toxics Rule (CTR) (CA)
2.1.3	Which are based on ecoregion/physiography	Temp and DO (MI, IN) ; None (RI,OR) ; None yet - possibly DO, TDS, Cl, SO_4, and nutrients (LA); Biocriteria[2] (MA) All except fecal coliform (AR) ; All except CTR, hot springs std (CA[3])
2.1.4	Are the WQS of surrounding states different for waterbodies that the states share in common?	Yes (OR/CA; AR/LA; MI/IN; MA)
	Which parameters?	Main differences are in classification or designated use (MA); LA and AR differ for chlorinated pesticides, turbidity, minerals (chloride, sulfate, TDS), Hg; Temperature (OR)
2.2.0	Which WQ parameters incorporate separate dry and wet weather conditions and criteria?	None (IN,RI,MA,AR,LA,OR,) ; Fecal coliform (MI); **Criteria** set for worst case (wet or dry) , wet and dry criteria set for specific sites (e.g. Newport Bay TMDL) (CA)
2.3.0	Have designated uses for waterbodies been scientifically demonstrated to be attainable?	No (IN,MA) ; Yes in most cases (MI,CA); Yes (AR) ; No response (OR, LA)

Number	Question	SUMMARY OF ALL EIGHT STATES
2.3.1	How were designated uses assigned to different waterbody types?	Based on support of warm or cold water biota (MI) ; Apply to all waters all places (IN); Rulemaking process (RI) EPA guidance, national drinking water goals, other state agencies, general public (MA) ; Data collection on reference streams (AR) ; CWA fishable/swimmable goals, others on actual use of goal (LA) ; Based on support of salmonid populations (OR) ; Statute, public hearings, field surveys, tributary rule (CA)
2.3.2	Were designated uses assigned to all waterbody types in the state or are some waterbody types exempted from specific uses such as drinking water?	DW apply only to intake waters (IN), Assigned to all (MI,RI,AR,MA); No response (CA); Some waterbodies are not covered by salmonid derived stds (OR)
2.3.3	Are fishable uses based on the existing fishery or on historical information?	Historical info (IN,MA,OR,CA); Both (MI); No response (RI); Existing fishery (AR)
2.3.4	How are threatened and endangered species incorporated in the designated waterbody uses?	Protected via Outstanding Resource Waters (ORW) classification (IN); Case-by-case basis (MI); No response (RI); General narrative criteria/antidegradation provisions and ORW (MA); Under the general narrative criteria and antidegradation provisions or through separate designation of ORW (MA); Ecologically sensitive designation (AR); Designated for fish/wild pops or habitat but not in listing decision or for T&E (LA, CA)
2.3.5	How many UAAs have been conducted on waterbodies in the state and primarily for which uses?	34, DO, ammonia - aq life (IN); 2, Ni, DO - wetlands (MI); None (RI); a couple, bacteria - primary contact (MA); 35, fish/swim, fisheries, primary contact, dw (AR); 46, DO (LA);> 20 in 30 years (CA)
2.3.6	How many water bodies may need or definitely need revised designated uses?	No response (MI, IN); Unknown (OR, MA); Few (AR,CA); Some (RI); Many would be eligible (LA)
2.4.0	What are the underlying scientific bases for linking designated uses to water quality criteria?	Assume waters can attain fish/swim uses (IN) Scientific literature on requirements of cold and warm water biota and human health tolerances (MI) Data from least disturbed waterbodies in each ecoregion adopted as criteria (AR) Chemical based on lab data, narratives on antidegradation, others site-specific analyses (CA); No response (OR, RI, MA, LA)
2.4.1	How were water quality criteria assigned to protect designated uses of waterbodies?	All waters designated for same uses, criteria set for uses (IN); They were designed (MI); National process and criteria (RI, MA, LA); Lit, studies, and professional knowledge (AR); Protection of salmonids (OR) Rulemaking, scientific evidence of regional boards (CA)

Number	Question	SUMMARY OF ALL EIGHT STATES
2.4.1.1	Were EPA national criteria used as the basis for state criteria?	Yes (RI, MA, LA) ; Often (CA); In some cases (AR); Yes except temp (OR)
2.4.1.2	Were the EPA national criteria modified prior to adoption by the state?	Yes, human health (RI); Yes, DO and temp (MA); In some cases, DO (AR);Yes, toxics - although not appropriate for the state (LA); Mostly not (CA)
2.4.1.3	What factors were considered in this modification?	Risk and exposure (RI); Site specific (MA); Local conditions (AR); Fish consumption and toxicity values (LA); dioxin, HH, risk values, WER (CA)
2.4.2	Do you consider the water quality criteria protective of the designated uses?	Yes (AR, LA, IN, OR, RI, MA); In most cases (MI) ; Strict for effluent, less definitive for ambient (CA)
2.4.3	How many site specific criteria have been developed for waterbodies in the state and primarily for which pollutants?	Metals: Cd, Cu, Pb, Ag, Zn (RI); DO and temp (MA); 30 to 40, minerals (Cl, SO4, TDS) (AR); 38, DO (LA); Temp (OR); More than 100 (CA)
2.4.4	How many water bodies may need or definitely need revised criteria?	Not sure (RI, MA); All might need it (AR); Many esp fecal (LA); May revise temp, toxics, nutrient, sed (OR); Clarify Cu and Ni stds (CA)
2.4.5	How would you develop water quality criteria that are protective of designated uses?	No response (IN);Updates to WQS (MI);Use site specific development policy (RI, MA); TMDL process (MA); Measure existing conditions supporting specific uses (AR) ; Conduct a peer reviewed and select a committee to decide (OR); Depends on how uses are defined (CA)
3.0 Determination of Impairment		
3.1	Was the 1997 EPA 303(d) guidance used to determine impairment?	Yes (LA, OR, IN, MA, AR) ; For streams (IN); Where applicable it applied (MI); with some adjustments (AR); and other EPA guidance (MA); No, used the 305b guidance (RI); No, it came out too late but probably consistent (CA)
3.2	Was this guidance modified in any way? If so how?	No (IN, MA, RI) different parameters have different levels of allowable exceedances (AR), Not Answered (LA, MI, OR, CA)
3.3	Does the state favor a policy that	Precautionary principle although may switch to degree of uncertainty in the future (IN, OR); half of each (RI); high degree of certainty (MA, AR, LA) No definitive policy, 1998 list used precautionary approach (CA)

Number	Question	SUMMARY OF ALL EIGHT STATES
	1. Requires a relatively high degree of certainty that a waterbody is truly impaired before it is listed, or	
	2. Applies a precautionary principle and allows a waterbody to be listed even if there is only modest evidence of impairment?	
3.4	Can a waterbody be listed for a single exceedance of a WQS?	yes (IN, LA); Yes, but depends on which standard (e.g., Hg) (CA); No (MI, RI, AR, OR); not unless other data clearly documents the problem (MA);
	If so, what portion of listed water bodies are listed on this basis?	None (MI); ca 25% (IN); <10% (CA); BPJ may be used when guidance isn't available (MA); acute toxicity > 1 (AR); not sure (LA)
3.5	Are there statistical criteria for listing that consider a minimum number of samples, percent exceedance, and mean of the pollutant concentration?	Not at this time (IN), process included in new TMDL listing document - Apr '02 (MI); No (RI); Yes (AR, MA, LA) ; All - do not use mean of pollutant (AR);
	What portion of listed water bodies are listed on this basis?	Unknown (LA); > 10% must show an exceedance before listing (OR); None (CA); No response (MI, IN, RI, MA, AR)
3.6	Are there listings of impairment or causes of impairment for which there are no criteria?	Yes, (IN, LA, MA, RI) for biological impairments (IN) evaluative assessments (no monitoring data) (LA) narrative criteria (CA); No (AR)
3.6.1	How are waterbodies treated if the WQS are exceeded but the designated uses are being achieved?	>1 sample violates a WQS = impaired (IN); If WQS are sufficiently shown to exceeded, then waterbody is listed; Listed as threatened (RI); generally listed with some exceptions (MA); Adjust assessment criteria (AR); listed impaired, but may need UAA or further sampling prior to TMDL development (LA); evaluate criteria & beneficial uses independently (OR); Needs lots of data, determined in TMDL process (CA)

Number	Question	SUMMARY OF ALL EIGHT STATES
3.6.2	Are spills, upsets, or bypasses during storm events considered as WQS exceedance and possible listing?	Spills: No (IN); CSOs & bypassing during a storm event: Yes (IN); Infrequent accidental bypasses/upsets: No (IN); All: on a case-by-case basis (MI); All: No (RI); Not for listing (AR); considered temporary and generally not listed (MA); Data is not separated out into dry and wet weather categories (OR); Yes, but not likely a single event (CA); Not sure (LA)
3.8	If biological criteria are used how were these criteria established?	Ranking using reference sites (IN); No response (RI); No (MA); not formal part of WQS (AR); (LA); Not using biological criteria, however, narrative includes concept of "don't mess up community structure" (CA); No response (OR, MI)
	Reference? Ranking? Other sites?	Narrative based on bioassessments, soon numeric biocriteria (IN);
3.8.1	Do these standards incorporate natural variability among areas?	Yes (MI, MA, CA); Ranges of metrics within ecoregions (AR), No response (RI, LA, OR, IN)
3.8.2	How are habitat considerations incorporated into the criteria?	Measured and compared to reference stream or reference condition (AR); No response (CA, IN, LA, MA, MI, OR, RI)
3.8.3	How were the biological criteria developed and tested?	Under development (IN); No criteria, RBP protocols used to determine impairment (MA); Field investigations (AR); No response (CA, LA, MI, OR, RI)
3.8.4	How are exceedances of biological criteria determined?	Index scores based on multimetric analysis fall below prescriptive values (IN); RBP II or II are used to determine if the segment supports its use (MA); IBI approach (AR); No response (CA, LA, MI, OR, RI)
3.9	For parameters incorporating both wet and dry weather values, how are wet weather data interpreted or evaluated?	Data are not separated into dry/wet weather (IN, RI,AR, LA, OR,); WQS exceedance of criteria (MI); on a case by case basis (CA); No response (MA)
3.10	Are listings based on a monitoring system?	Yes (IN, MI, RI, MA,AR, LA, CA, OR)
3.10.1	What design was used for the monitoring program?	5 year watershed monitoring cycle (MA); Statewide monitoring plan under development (CA)
	Fixed station?	Yes (IN, MI, RI, MA, AR, LA, OR, CA)
	Probabilistic?	Yes (IN, RI, OR, CA) ; No (MI, AR) No but considering for 305b purposes (MA)

Number	Question	SUMMARY OF ALL EIGHT STATES
	Problem-based (e.g. NPDES dischargers)?	Yes (IN, RI, AR); Most monitoring is this (CA)
	Based on historical location?	Yes (IN, RI, AR, MA);
	Other reasons (specific)	Federal grant opportunities (OR); Public requests (MA); fixed sampling by USGS, intensive surveys, special studies, volunteer monitoring programs (RI)
3.10.2	Is the design for the 305b assessment the same as used for listing?	305b design is the same used for listing waters in the 303d report (IN); Yes (MI, RI, MA, LA); Use fixed station network and probabilistic monitoring data (OR); No (CA)
	Are there plans for modifying the monitoring network or evaluative procedure approaches?	CALM (IN); Yes (RI); Adopting probabilistic approach as one of the options for 305b purposes (MA); Now a rotating basin approach on a 5-year cycle, and maybe new assessment methods (LA); Network is under constant evaluation for modification (AR)
3.11	Are listings based on evaluative procedures?	Yes (AR), No (IN, RI); Generally not (MI) DEP sometimes uses data and analysis by other agencies and sources (MA); Historically yes, but in 1998 LDEQ discontinued evaluative assessment (LA); Evaluation of habitat condition and a determination of beneficial use impairment (OR); Yes, weight of evidence (CA)
3.11.1	To what parameters are evaluative procedures applied?	Mercury, maybe, PCBs (MI); Chemical and biological (RI); Bacteria and metals or organics in fish tissue (MA); In principle, evaluative procedures can be applied to all parameters (AR); Evaluations were used for habitat modification, flow modification and sedimentation (OR); All (LA, CA)
3.11.2	What determines the choice of evaluative procedures over monitoring?	Extensive distribution due to atmospheric loading of a given pollutant (Hg, PCBs, chlordane) (OR); Availability of resources (RI); Conditions similar to monitored segment; if unmonitored and monitored are similar, evaluation procedures are used for attainment (AR); Evaluative procedures were used where no numeric criteria were available (OR)
3.11.3	Were evaluation procedures tested and verified?	Yes (MA), Yes by EPA (RI); Used BPJ but did not use follow up data collection (AR); No, but questionnaires were sent with available monitoring data. Field coordinators were given guidance regarding the evaluative assessment (LA). No (CA, OR)
3.11.4	Are evaluation procedures used as part of the 305(b) evaluation process?	Yes (IN, MA, AR, CA, OR); The 305b review process non-attainment listing is used in augmenting the 303d TMDL list (MI); Not for the 2000 cycle (RI); Yes they were until 1998, but are no longer (LA).

Number	Question	SUMMARY OF ALL EIGHT STATES
3.12	Were data from 319 projects used to evaluate impairment?	Minimal, not much data from 319 projects were available for the 1998 List (IN); Where available (MA); Yes (CA); No, 319 projects are developed from the evaluation process (AR); No (LA, RI); Data submitted during the public notice period was used (OR)
3.13	Are narrative criteria used to list waterbodies?	Yes (RI, IN, MI, MA, AR, OR, CA); No (LA)
3.13.1	How are narrative criteria applied if used to list waterbodies?	Bioassessment results (IN); RBP methodologies (MA); Limits phosphorus discharge to prevent water degradation (MI); The phrase "no toxics in toxic amounts" is used RI); narrative criteria must be supported by data for an associated numeric standard (AR); Evaluate data and reports to find evidence of use impairment (OR); No (LA)
3.13.2	Which narrative criteria are violated most often?	Biological (IN, RI) oil and grease (IN); Phosphorus (MI); Nutrients, toxicity, and aesthetics (MA); None (AR); Sediment, temperature, aquatic community structure (CA)
3.13.3	Do narrative criteria used for listing have objective evaluative procedures and approaches for ensuring designated uses are impaired and consistently implemented among water bodies?	Yes (RI), Yes, MIBI, IBI, QHEI (IN); Yes and No (MI); Yes, in most cases, but BPJ may be used given the amount and type of data (MA); No (OR, CA); None used to list impaired, may list waters of concern (AR)
4.0 Data Quality		
4.1	Are there minimum data quality requirements for data that are used in impairment determination?	Yes - all states except MI (no response) and CA (WQ data collected under QA plan depending on the study)
4.2	What quality assurance/quality control procedures are used to ensure the data are scientifically credible?	Documented in QAPP for all states except MI (no response); OR (DEQ requests QA/QC data be available)
4.3	How old are the data used for listing?	Typically, 1-5 years except MI (no response); 5-10y (OR); and all (CA)
	1-2 yr 3-5 yr 5-10 yr >10 yr	

Number	Question	SUMMARY OF ALL EIGHT STATES
4.4	What is your estimate of the reliability of the data used for listing and delisting?	Good (RI), fairly reliable (IN), very reliable (AR, CA, LA), former data (>5 yrs old) may be questionable (MA) and no response (MI, OR)
4.5	How could the reliability of the data be improved?	More data (IN, AR); Require submission of QA/QC data (OR); Data requirements that were instituted (MA); More frequent sampling, more parameters(RI); No response (MI,LA)
5.0 Delisting		
5.1	How many waterbodies have been delisted since 1995?	3 during '96-'98 (MI), 42 in 1998 (RI), '68 (MA), >100 since '99 (LA), 476 miles (OR), none (AR), no response (IN)
5.2	Was information for delisting obtained from a statistically designed monitoring network or based on verified and validated evaluative procedures?	Monitoring data (MI,RI,LA, MA), listing and delisting are the same (monitoring data and evaluative procedures) (OR), evaluative procedures (IN), New data; controls implemented, anecdotal info, but not evaluative procedures (MA); No response (AR)
5.3	What are your approaches or criteria for delisting waters?	
	Same as listing	Yes (IN,MA), Similar (MI, CA)
	New or additional data	Yes (IN,MI,RI,MA,LA,OR,CA)
	Completed TMDL	No (IN) No - but should be allowed by EPA (MA), Yes (MI,RI,OR,CA)
	Statistical significant trend indicating recovery	Ideally where nutrient TMDL is implemented (MI) No (IN, CA)
	Redefining water quality standards (criteria, designated uses, anti-degradation)	Yes (MA,LA,CA, AR)
	Other	Control measures (e.g., permits, watershed mgmt plans) (CA)
5.4	Are trend analysis or power analysis used to determine if a waterbody can be delisted?	No (IN,MI,RI,MA,AR,LA,OR) [1]; No response (CA)

Number	Question	SUMMARY OF ALL EIGHT STATES
5.5	If there is reasonable assurance the implementation plan will result in the recovery of the waterbody, can it be delisted?	No (IN,MA,AR,LA); Delisted upon TMDL, approval by EPA, source elimination or remediation (MI); Not sure (RI); No response (OR,CA)
5.6	Are waters delisted only when a new 303(d) list is submitted to EPA or are they delisted as the TMDLs or other criteria for delisting are satisfied?	New 303(d) list is required (IN,MI,RI,MA,OR); Other criteria for delisting are satisfied (AR) ; New data are available/new re-assessment (LA); No response (CA)
5.7	What criteria are used to delist waterbodies that have been listed based on narrative criteria?	Removal of source of impairment (MI); Data needs to be more definitive or show the problem no longer exists (MA, RI, OR, CA); No response (AR,LA, IN); Approved TMDL by EPA (OR, CA); Objectives are revised, control measures implemented, or use id de-designated after UAA approval (CA)
8.0 Comments and Recommendations Regarding TMDLs		
8.3.1	How could the listing process be improved?	Insure flexibility in scheduling waterbodies (IN); EPA to provide clearer & practical guidance, consistent between regions (RI); Criteria based upon high quality data, only list truly impaired waters, eliminate public comment and timelines for comment prior to development and submittal (MA); Have each state develop listing/delisting methodologies and criteria, as is currently being done. Accept that a TMDL process is not the only way to delist a waterbody from the TMDL listing (MI); Improve assessment criteria. States need flexibility in how they bring information together to determine impairment. Knowledge of local conditions is crucial in determining which information should be brought to bear and which information should be given the greatest weight in the decision process. Only the states are in a position to do this (AR); The listing process could be improved by: 1) Identifying/defining the pollutants for which a waterbody can be impaired, That is, use only pollutants that are appropriate in the TMDL context. 2) restricting listing on 303(d) to waterbodies for which there is water quality monitoring data to show that a standard is not being met and pollutants that are suitable to TMDL calculation, 3) eliminate the evaluative assessment as a basis for listing on the 303(d) list, 4) eliminate the threatened and partial support assessment categories (water quality either meets standards or it doesn't), 5) allow for alternative listing of waterbodies for which a standard may be inappropriate or unattainable and in need of a UAA or which do not meet a standard due to natural conditions (LA)

Number	Question	SUMMARY OF ALL EIGHT STATES
8.3.2	How could the delisting process be improved?	Unsure – no problems requesting delisting from the 1996 to 1998 (IN); EPA to provide clearer & practical guidance, consistent between regions (RI); Allow de-listing based upon new data, completion of a TMDL, or demonstration that the original listing was incorrect or based upon anecdotal or limited information (MA); Noted that there is not a process (AR); Delisting can be accomplished by removal of sources of pollutants (stressor) whether point source, nonpoint source, atmospheric or in-place sources (MI); The delisting process could be improved by: 1) Requiring equivalent data or information for delisting that was used to list, it shouldn't take more data to delist than was used to list a waterbody and if an evaluative assessment was the basis for listing, an evaluative assessment should be sufficient for delisting, 2) expedite EPA review and approval (do not involve attorneys in the review of technical submittals) or eliminate EPA approval authority (LA).

NOTES:

[1] All parameters for Michigan State are assumed to be uniform except Temperature and DO, based on the answer

[2] Parameter under development

[3] CTR - California Toxics Rule

[1] For question 5.4, not deemed necessary by MI; most cases insufficient data to make this determination in MA; AR acknowledged it could be

REFERENCES

Arkansas Department of Pollution Control and Ecology (ADPCE). 1987. Physical chemical and biological characteristics of least-disturbed reference streams in Arkansas' ecoregion. Vol. I and II. ADPCE, Little Rock, AR.

Chapman, P.M. 2000. Whole effluent toxicity testing-usefulness, level of protection, and risk assessment. *Environ. Toxicol. Chem.* 19:3-24

Chapman, P.M. 1986. Sediment quality criteria from the sediment quality triad: An example. *Environ. Toxicol. Chem.* 5:956-965.

de Vlaming, V., and T.J. Norberg-King. 1999. A review of single species toxicity tests: Are the tests reliable predictors of aquatic ecosystem community response? EPA/600/R-97/114. EPA Office of Research and Development, Duluth, MN.

Diamond, J., and C. Daley. 2000. What is the relationship between whole effluent toxicity and instream biological condition? *Environ. Toxicol. Chem.* 19: 158-168.

Dickson, K.L, W.T. Waller, J.H. Kennedy, and L.P. Ammann. 1992. Assessing the relationship between ambient toxicity and instream biological response. *Environ. Toxicol. Chem.* 11:1307-1322.

Florida Department of Environmental Protection (FDEP). 2002. Florida Pilot Mercury Total Maximum Daily Load (TMDL) Study: An Assessment of the Effects of Changing Mercury Emission on Aquatic Cycling and Bioaccumulation in the Everglades. Integrated Study Report. FDEP and EPA, Tallahassee, FL.

FWQC/AFBF/AMSA. 2002. Preparation of Integrated water quality monitoring and assessment reports: Recommendations for Clean Water Act 303(d) and 305(b) methodologies and reporting. Federal Water Quality Coalition, American Farm Bureau Federation and the Association of Metropolitan Sewerage Agencies. Prepared by Barnes and Thornburg, Chicago, and Parsons, Austin TX, March 11, 2002.

GAO. 2000. Water Quality — Key EPA and State Decisions Limited by Inconsistent and Incomplete Data. GAO/RCED-00-54. Washington, DC, U.S. General Accounting Office.

GAO. 2002 Water Quality — Inconsistent state approaches complicate nation's efforts to identify its most polluted waters. GAO-02-186. U.S. General Accounting Office, Washington DC.

Hall, L.W. and J.M. Giddings, 2000. The need for multiple lines of evidence for predicting site specific ecological effects. *Hum. Ecol. Risk Assess.* 6:679-710.

Harmancioglu, N.B., V. Singh, O. Fistikoglu, S.D. Ozkul, M.N. Alpaslan. 1999. *Water Quality Monitoring Network Design.* Kluwer Academic Publishers.

Heber, M.A., D.K. Reed-Junkins, YY Davies. 1996. USEPA's whole effluent toxcitiy testing program: A national regulatory perspective. Pg. 9-15 In: D.R. Groeth, K.L. Dickson, D.K. Reed-Junkins eds. *Predicting Instream Effects From WET Tests: An Evaluation of Methods and Prediction of Receiving Stream Impacts.* SETAC Press, Pensacola, FL.

Hughes, R.M. 1995. Defining acceptable biological status by comparing with reference conditions. P 15-30 In: Davis, W.S. and T.P. Simon. Eds. *Biological Assessment and Criteria: Tools for Water Resource Planning and Decision Making.* Lewis Publishers, Boca Raton, FL.

Hughes, R.M., and D.P. Larsen. 1988. Ecoregions: an approach to surface water protection. *JWPCF* 60: 486-493

Hughes, R.M., S.A. Heiskary, W.J. Matthews, and C.O. Yoder. 1994. Use of ecoregions in biological monitoring. Pg. 125-151. In: S.L. Loeb and A. Spacie. Eds. *Biological Monitoring of Aquatic Systems.* Lewis Publishers, Chelsea, MI.

Larsen, D.P., K.W. Thornton, N.S. Urquhart, and S.G. Paulsen. 1994. The role of sample surveys for monitoring the condition of the Nation's lakes. *Environ. Monit. and Assess.* 32: 101-134.

LDEQ. 2000. 2000 305(b) Post III: Surface Water Assessment. Louisiana Department of Environmental Quality, Baton Rouge, LA.

Lin, P., D. Meeter, and X. Niu. 2000. A nonparametric procedure for listing and delisting impaired waters based on criterion exceedances. Tech. Rept. Submitted to Florida Department of Environmental Protection, Contract No. LAB015; Department of Statistics, Florida State Univ., Tallahassee, FL, October 2000.

Macauley, J.M., J.K. Summers, and V.D. Engle. 1999. Estimating the ecological condition of the estuaries of the Gulf of Mexico. *Environ. Monit. and Assess.* 57:59-83.

National Research Council (NRC). 2001. *Assessing the TMDL Approach to Water Quality Management.* Washington, DC, National Academy Press.

Olsen, A.R., J. Sedransk, D. Edwards, C.A. Gotway, W. Liggett, S.L. Rathburn, K.H. Reckhow, and L.J. Young. 1999. Statistical issues for monitoring ecological and natural resources in the United States. *Environ. Monit. and Assess.* 54: 1-45.

Omernik, J.M. and G.E. Griffith. 1991. Ecological regions vs hydrologic units: frameworks for managing water quality. *J. Soil. Water. Conserv.* 46: 334-340.

Omernik, J.M., 1995. Ecoregions: A Spatial Framework for Environmental Management. Pg. 49-61 In: Davis, W.S. and T.P. Simon. 1995. *Biological Assessment and Criteria: Tools for Water Resources Planning and Decision Making.* Lewis Publishers. Boca Raton, FL.

Smith, E.P., K. Ye, C. Hughes, and L. Shabman. 2001. Statistical assessments of violations of WQS under Section 303(d) of the Clean Water Act. *Env. Sci. Technol.* 35:606-612.

Stevens, D.L. Jr. 1994 Implementation of a national monitoring program. *Environ. Manage.* 42: 1-29.

Thornton, K.W. and C. Creager. 2001. Chapter 6-Watershed Management, In: Holdren, C., W. Jones, and J. Taggart Eds. *Managing Lakes and Reservoirs.* 3rd Edition. N. Am. Lake Mgmt. Soc., Madision, WI.

Texas National Resource Conservation Commission (TNRCC). 2002. Guidance for answering Texas surface and finishes drinking water quality data. TNRCC, October 16, 2001.

U.S. Environmental Protection Agency. 1990. Biological criteria: National program guidance for surface waters. EPA-440/5-90-004. Office of Water.

U.S. Environmental Protection Agency. 1991. Guidance for Water Quality-Based Decisions: The TMDL Process. Office of Water, Washington, DC. EPA 440/4-91-001. April 1991.

U.S. Environmental Protection Agency. 1994a. Guidance for the Data Quality Objectives Process EPA QA/R-4. Office of Research and Development, Washington, DC. EPA/600/R-96/055, September 1994.

U.S. Environmental Protection Agency. 1994b. Interim guidance on the determination of and use of water-effect ratios for metals. EPA-823-B-94-001, Office of Science and Technology.

U.S. Environmental Protection Agency. 1997. Guidelines for preparation of comprehensive state water quality assessments (305(b) Reports) and electronic updates. EPA 841-B-97-002A.

U.S. Environmental Protection Agency. July 1998a. Report of the Federal Advisory Committee on the Total Maximum Daily Load (TMDL) Program. Office of the Administrator, Washington, DC.

U.S. Environmental Protection Agency. 1998b. 1998 TMDL Tracking Database. http://www.epa.gov/owow/tmdl/trcksys.html (2001 Version), Office of Water, Washington, DC.

U.S. Environmental Protection Agency. 1999. Draft guidance for water quality-based decisions: The TMDL process (2nd Edition), Office of Water, Washington, DC.

U.S. Environmental Protection Agency. 2000a. Stressor identification guidance document. EPA/822/B-00/025, Office of Research and Development, Washington, DC.

U.S. Environmental Protection Agency. 2000b. EPA review of 2000 Section 303(d) lists. Memorandum, Robert H. Wayland, Director, Office of Wetlands, Oceans, and Watersheds, April 28, 2000.

U.S. Environmental Protection Agency. 2001a. EPA requirements for quality assurance project plans EPA QA/R5, EPA/240/B-1/003, US Environmental Protection Agency, Office of Environmental Information, Washington, DC, March 2001.

U.S. Environmental Protection Agency. 2001b. 2002 Integrated water quality monitoring and assessment report guidance. Memorandum from Robert H Wayland, Director, Office of Wetlands, Oceans and Watersheds to EPA Regional Water Management Directors, EPA Regional Science and Technology Directors, and State, Territory and Authorized Tribe Water Quality Program Directors.

U.S. Environmental Protection Agency. 2001c. Clarification of the use of biological data and information in the 2002 integrated water quality monitoring and assessment report guidance. Memorandum from Robert H Wayland, Director, Office of Wetlands, Oceans and Watersheds to EPA Regional Water Management Directors, EPA Regional Science and Technology Directors, and State, Territory and Authorized Tribe Water Quality Program Directors, March 26, 2002.

U.S. Environmental Protection Agency. 2002. Consolidated Assessment and Listing Methodology: Toward a Compendium of Best Practices. Draft for State Review and Comment. Prepared by EPA Office of Wetlands, Oceans and Watersheds, April 20, 2001.

Waller, W.T., L.P. Ammanns, W.J. Birge, K.L. Dickson, P.B. Dorn, N.E. LeBlanc, D.J. Mouint, B.R. Parkhurst, H.R. Preston, S.C. Schimmel, A. Spacie, and G.B. Thursby. 1996. Predicting instream effects from WET tests. p. 271-286 In: D.R. Groethe, K.L. Dickson, D.K. Reed-Junkins, Eds. *Predicting Instream Effects From WET Tests: An Evaluation of Methods and Prediction of Receiving Stream Impacts.* SETAC Press, Pensacola, FL.

Ward, R.C., J.C. Loftis, and G.B. McBride. 1997. *Design of Water Quality Monitoring Systems.* Van Nostrand Reinhold.

Water Environment Research Foundation (WERF). 1997. Use attainability analysis: A comprehensive UAA technical reference. Final Report, Project 91-NPS-1. WERF, Alexandria, VA.

Yoder, C.O., and E.T. Rankin. 1995a. Biological criteria program development and implementation in Ohio. Pg. 109-144. In: Davis, W.S. and T.P. Simon, Eds. 1995. *Biological Assessment and Criteria. Tools for Water Resources Planning and Decision Making.* Lewis Publishers, Boca Raton, FL.

Yoder, C.O., and E.T. Rankin. 1995b. Biological response signatures and the area of degradation value: new tools for interpreting multimetric data. Pp. 263-286. In: Davis, W.S. and T.P. Simon, Eds. 1995. *Biological Assessment Criteria: Tools for Water Resources Planning and Decision Making.* Lewis Publishers, Boca Raton, FL.

2003 WERF SUBSCRIBERS

Wastewater Utility

Alabama
Montgomery Water Works & Sanitary Sewer Board

Alaska
Anchorage Water & Wastewater Utility

Arizona
Gila Resources
Glendale, City of, Utilities Department
Mesa, City of
Phoenix Water Services Department
Pima County Wastewater Management

Arkansas
Little Rock Wastewater Utility

California
Calaveras County Water District
Central Contra Costa Sanitary District
Contra Costa Water District
Crestline Sanitation District
Delta Diablo Sanitation District
Dublin San Ramon Services District
East Bay Municipal Utility District
El Dorado Irrigation District
Escondido, City of
Fairfield-Suisun Sewer District
Irvine Ranch Water District
Las Virgenes Municipal Water District
Lodi, City of
Los Angeles, City of
Los Angeles County, Sanitation Districts of
Napa Sanitation District
Orange County Sanitation District
Palo Alto, City of
Riverside, City of
Sacramento Regional County Sanitation District
San Diego Metropolitan Wastewater Department, City of
San Francisco, City & County of
San Jose, City of
Santa Barbara, City of
Santa Rosa, City of
South Bayside System Authority
South Orange County Wastewater Authority
Union Sanitary District
West Valley Sanitation District

Colorado
Boulder, City of
Colorado Springs, City of
Littleton/Englewood Water Pollution Control Plant
Metro Wastewater Reclamation District, Denver

Connecticut
New Haven, City of, WPCA

District of Columbia
District of Columbia Water & Sewer Authority

Florida
Broward, County of
Fort Lauderdale, City of
Gainesville Regional Utilities
JEA
Kissimmee, City of, Department of Water Resources
Miami-Dade Water & Sewer Authority
Orange County Utilities Department
Orlando, City of
Reedy Creek Improvement District
Seminole County Environmental Services
St. Petersburg, City of
Stuart Public Utilities
Tallahassee, City of
Tampa, City of
West Palm Beach, City of

Georgia
Atlanta Wastewater Services
Augusta, City of
Clayton County Water Authority
Cobb County Water System
Columbus Water Works
Fulton County
Gwinnett County Department of Public Utilities
Macon Water Authority
Savannah, City of

Hawaii
Honolulu, City and County of

Idaho
Pocatello, City of

Illinois
American Bottoms Wastewater Treatment Plant
Dupage, County of
Greater Peoria Sanitary District
Kankakee River Metropolitan Agency
Metropolitan Water Reclamation District of Greater Chicago
Wheaton Sanitary District

Indiana
Fort Wayne, City of
Indianapolis, City of

Iowa
Ames, City of
Des Moines Metro Wastewater Reclamation Authority

Kansas
Johnson County Unified Wastewater Districts
Unified Government of Wyandotte County/Kansas City, City of

Kentucky
Louisville & Jefferson County Metropolitan Sewer District

Louisiana
Sewerage & Water Board of New Orleans

Maine
Bangor, City of
Portland Water District

Maryland
Anne Arundel County Bureau of Utility Operations
Howard County Department of Public Works
Washington Suburban Sanitary Commission

Massachusetts
Boston Water & Sewer Commission
Upper Blackstone Water Pollution Abatement District

Michigan
Ann Arbor, City of
Detroit, City of
Grand Rapids, City of
Holland Board of Public Works
Lansing, City of
Owosso Mid-Shiawassee County WWTP
Saginaw, City of
Wayne County Department of Environment
Wyoming, City of

Missouri
Independence, City of
Kansas City Missouri Water Services Department
Little Blue Valley Sewer District
Metropolitan St. Louis Sewer District

Nebraska
Lincoln Wastewater System

Nevada
Henderson, City of

New Jersey
Passaic Valley Sewerage Commissioners

New York
New York City Department of Environmental Protection

North Carolina
Cary, Town of
Charlotte/Mecklenburg Utilities
Durham, City of
Metropolitan Sewerage District of Buncombe County
Orange Water & Sewer Authority

Ohio
Akron, City of
Butler County Department of Environmental Services
Columbus, City of
Metropolitan Sewer District of Greater Cincinnati
Northeast Ohio Regional Sewer District

Oklahoma
Tulsa, City of

Oregon

Clean Water Services

Eugene/Springfield Water Pollution Control

Water Environment Services

Pennsylvania

Philadelphia, City of

University Area Joint Authority, State College

South Carolina

Charleston Commissioners of Public Works

Mount Pleasant Waterworks & Sewer Commission

Spartanburg Sanitary Sewer District

Tennessee

Cleveland, City of

Knoxville Utilities Board

Murfreesboro Water & Sewer Department

Nashville Metro Water Services

Texas

Amarillo, City of

Austin, City of

Dallas Water Utilities

Denton, City of

El Paso Water Utilities

Fort Worth, City of

Gulf Coast Waste Disposal Authority

Houston, City of

San Antonio Water System

Trinity River Authority

Utah

Salt Lake City Corporation

Virginia

Alexandria Sanitation Authority

Arlington, County of

Fairfax County Virginia

Hampton Roads Sanitation District

Henrico, County of

Hopewell Regional Wastewater Treatment Facility

Loudoun County Sanitation Authority

Lynchburg Regional WWTP

Prince William County Service Authority

Richmond, City of

Washington

Edmonds, City of

Everett, City of

King County Department of Natural Resources

Seattle Public Utilities

Wisconsin

Green Bay Metro Sewerage District

Madison Metropolitan Sewerage District

Milwaukee Metropolitan Sewerage District

Racine, City of

Sheboygan Regional Wastewater Treatment

Wausau Water Works

Australia

Sydney Water Corporation

Water Corporation of Western Australia

Canada

Toronto, City of, Ontario

Mexico

Servicios de Agua y Drenaje de Monterrey, I.P.D.

United Kingdom

Yorkshire Water Services Limited

Stormwater Utility

California

Monterey, City of†

Santa Rosa, City of†

Colorado

Boulder, City of†

Georgia

Griffin, City of†

Kansas

Overland Park, City of†

Kentucky

Louisville & Jefferson County Metropolitan Sewer District†

Maine

Portland Water District†

North Carolina

Cary, Town of†

Charlotte, City of/Mecklenburg County†

Pennsylvania

Philadelphia, City of†

State

Fresno Metropolitan Flood Control District, Calif.†

Urban Drainage & Flood Control District, Colo.†

Corporate

ADS Environmental Services

The ADVENT Group Inc.

Alan Plummer & Associates

Alden Research Laboratory

Alpine Technology Inc.

Aquateam–Norwegian Water Technology Centre A/S

BaySaver Inc.†

BioVir Inc.

Black & Veatch

Boyle Engineering Corporation

BPR CSO

Brown & Caldwell

Burns & McDonnell

CABE Associates Inc.

The Cadmus Group

Camp Dresser & McKee Inc.

Carollo Engineers Inc.

Carpenter Environmental Associates Inc.

CDS Technologies Inc.

Chemtrac Systems Inc.

CH2M HILL

Clancy Environmental Consultants Inc.

Damon S. Williams Associates, LLC

Earth Tech Inc.

Ecolab Water Care Services

EMA Inc.

The Eshelman Company Inc.

Finkbeiner, Pettis, & Strout Inc.

Frontier Geosciences Inc.

ftn Associates Inc.

Gannett Fleming Inc.

GE Betz

Golder Associates Inc.

Greeley and Hansen LLC

The HACH Company

Hazen & Sawyer, P.C.

HDR Engineering Inc.

HNTB Corporation

HydroQual Inc.

Insituform Technologies Inc.

Institute for Environmental Technology & Industry, Korea

Jacobson Helgoth Consultants Inc.

Jason Consultants Inc.

Jordan, Jones, & Goulding Inc.

KCI Technologies Inc.

Kelly & Weaver, P.C.

Kennedy/Jenks Consultants

Komline Sanderson Engineering Corporation

Lawler, Matusky & Skelly Engineers, LLP

Limno-Tech Inc.

Lombardo Associates Inc.

Malcolm Pirnie Inc.

McKim & Creed

MWH

New England Organics

Odor & Corrosion Technology Consultants Inc. (OCTC)

ONDEO Degremont Inc.

PA Government Services Inc.

Parametrix Inc.

Parsons

Post, Buckley, Schuh & Jernigan

The RETEC Group

R.M. Towill Corporation

Ross & Associates Ltd.

Royce Technologies

SAIC Maritime Technical Group

Sear-Brown

Sierra Environmental Services

SYNAGRO

Tetra Tech Inc.

Trojan Technologies Inc.

URS Corporation

USFilter NATC

Vortechnics Inc.†

Wade-Trim Inc.

Weston Solutions Inc.

Woodard & Curran

WRc/D&B, LLC

WWETCO, LLC†

Industry

American Electric Power

BP

ChevronTexaco Energy Research & Technology Company

The Coca-Cola Company

Dow Chemical Company

DuPont Company

Eastman Chemical Company

Eastman Kodak Company

Merck & Company Inc.

ONDEO Services

Procter & Gamble Company

Severn Trent Services Inc.

Shell Global Solutions

Thames Water Plc

United Water Services LLC

† Charter Stormwater Subscribers

Note: List as of 1/03

WERF Product Order Form

As a benefit of joining the Water Environment Research Foundation, subscribers are entitled to receive one complimentary copy of all final reports and other products. Additional copies are available at cost (usually $10). To order your complimentary copy of a report, please write "free" in the unit price column. WERF keeps track of all orders. If the charge differs from what is shown here, we will call to confirm the total before processing.

Name Title

Organization

Address

City State Zip Code Country

Phone Fax Email

Stock #	Product	Quantity	Unit Price	Total
			Postage & Handling	
			VA Residents Add 4.5% Sales Tax	
			Canadian Residents Add 7% GST	
			TOTAL	

Method of Payment:

(All orders must be prepaid.)

❑ Check or Money Order Enclosed

❑ Visa ❑ Mastercard ❑ American Express

Account No. Exp. Date

Signature

Shipping & Handling:

Amount of Order	United States	Canada & Mexico	All Others
Up to but not more than:	Add:	Add:	Add:
$20.00	$5.00*	$8.00	50% of amount
30.00	5.50	8.00	40% of amount
40.00	6.00	8.00	
50.00	6.50	14.00	
60.00	7.00	14.00	
80.00	8.00	14.00	
100.00	10.00	21.00	
150.00	12.50	28.00	
200.00	15.00	35.00	
More than $200.00	Add 20% of order	Add 20% of order	

*minimum amount for all orders

Note: Please make checks payable to the Water Environment Research Foundation.

To Order (Subscribers Only):

Log on to www.werf.org and click on the "Product Catalog."

Phone: (703) 684-2470
Fax: (703) 299-0742.

WERF
Attn: Subscriber Services
601 Wythe Street
Alexandria, VA 22314-1994

To Order (Non-Subscribers):

Non-subscribers may be able to order WERF publications either through WEF (www.wef.org) or IWAP (www.iwapublishing.com). Visit WERF's website at www.werf.org for details.